Seismic Resistant Design
and
Technology

Seismic Resistant Design and Technology

Dentcho Ivanov

New Jersey Institute of Technology
University Heights
Newark, New Jersey

CRC Press

Taylor & Francis Group

Boca Raton London New York

CRC Press is an imprint of the
Taylor & Francis Group, an **informa** business

A SCIENCE PUBLISHERS BOOK

CRC Press
Taylor & Francis Group
6000 Broken Sound Parkway NW, Suite 300
Boca Raton, FL 33487-2742

First issued in paperback 2020

© 2015 by Taylor & Francis Group, LLC
CRC Press is an imprint of Taylor & Francis Group, an Informa business

No claim to original U.S. Government works

ISBN-13: 978-1-4987-0536-3 (hbk)
ISBN-13: 978-0-367-73811-2 (pbk)

Library of Congress Cataloging-in-Publication Data

Ivanov, Dentcho.
    Seismic resistant design and technology / author, Dentcho Ivanov.
        pages cm
    Includes bibliographical references and index.
    ISBN 978-1-4987-0536-3 (hardcover : alk. paper) 1. Earthquake resistant design. I.
    Title.

    TA658.44.I93 2015
    693.8'52--dc23   2015018286

Visit the Taylor & Francis Web site at
http://www.taylorandfrancis.com

and the CRC Press Web site at
http://www.crcpress.com

Preface

Seismic waves are nonlinear bulk and surface elastic waves that propagate in the nonlinear, dissipative, dispersive, and heterogeneous medium of the Earth's body and on its surface. Earthquakes cause deaths and destructions all around the world. The main strategy for fighting the destructive seismic power has been designing reinforced building constructions and hoping that they will be able to withstand the tremor. Enormous progress has been made in this field. However earthquakes continue to be a big problem because of their unpredictability and diverse characteristics of the seismic waves involved. Especially destructive are some types of surface seismic waves.

In this book a different strategy for fighting the destructive power of earthquakes is discussed—developing seismic shields of metamaterials around cities that are able to deflect or dissipate the power of on-coming seismic waves.

Earthquakes even of moderate magnitude often cause major damages to cities. On one hand this is due to nonlinear phenomena such as growth of harmonics and waves of combination frequencies, parametric amplification, self-modulation, resonance effects, energy trapping, self-focusing, and wave-wave interactions and on the other hand to the strong nonlinearity and dispersion of the rock medium of propagation. The governing equations describing the wave propagation are nonlinear differential equations that can be solved only numerically assuming suitable approximations. One of most used approximation is to neglect all quadratic, cubic, and higher-order terms in the Piola-Kirchhoff stress tensor and keep only the first-order term. This approximation is the basis of the linear theory of elasticity. Using the elegant tools of Fourier analysis exact solutions of governing equations can be provided assuming infinitesimal wave amplitudes, invariance of the waveform during the propagation, and linear elastic response of the rock. Classical seismology has used continuum mechanics, linear elasticity theory, and general theory of scattering to describe the seismic wave behavior. However, seismic waves are finite-amplitude elastic waves and the elastic response of rocks is highly nonlinear. If the world were linear, there would be no earthquakes with characteristics we know.

The origin of seismic waves is discussed in Chapter 1. It is shown that nonlinearity causes growth of higher harmonics and waves of combination frequencies, which leads to 'seismic beat' and self-modulation. This explains the origin of 'coda waves' as high-frequency 'carrier waves' modulated by low-frequency body or surface elastic waves.

In Chapter 2 linear and nonlinear body waves in non-dispersive, dispersive, and heterogeneous media are discussed. Nonlinear phenomena of parametric amplification, elastic solitons, and stress-induced anisotropy are analyzed.

Linear and nonlinear surface elastic waves of Rayleigh, Love, Stoneley, Lamb, and Sezawa are discussed in Chapter 3 including body-to-surface and surface-to-body wave conversions, parametric amplification of Rayleigh waves, solitary elastic waves, solitons, skimming surface waves, and wave-wave interactions.

Experimental modelling and seismic shields for deflecting, scattering, and conversion of seismic waves are discussed in Chapter 4. Tools of physical acoustics, acoustoelectronics, acoustooptics and nondestructive testing of materials are shown to be suitable techniques for studying mini quakes in isotropic, anisotropic, and heterogeneous media. These techniques have been used successfully for developing a wide range of micro-electro-mechanical systems (MEMS). We will show that they can be applied also to seismic waves propagating in the giga-electro-mechanical systems (GEMS) of the Earth. Seismic shields of metamaterials can be experimentally modelled using pulse-laser generation of nonlinear body and surface elastic waves and laser-probe detection of their parameters.

Dentcho Ivanov

Contents

Introduction

On October 17, 1989 the Loma Prieta earthquake of moment magnitude 6.9 ('strong') on the Gutenberg-Richter scale and IX ('violent') on the Mercalli scale, with 19 km deep focus, and epicenter located 127 km south of San Francisco Bay Area in Nisene Marks State Park, California, caused the deaths of 63 people, injured another 3,757, and destroyed building and bridges in the Bay Area (Glough 1994).

On January 17, 1994 the Northridge earthquake of moment magnitude 6.7 ('strong') on the Gutenberg-Richter scale and IX ('violent') on the Mercalli scale, with 18 km deep focus, and epicenter in the middle of the densely populated neighborhood of Reseda in the San Fernando Valley, north-central Los Angeles caused the deaths of 57 people and injured another 8,700. During the Northridge earthquake the fastest seismic wave peak velocity of 6.59 m/sec and the highest ground acceleration of 16.7 m/sec^2 ever in North America were measured (Stover 1993).

The two earthquakes had almost the same parameters on Gutenberg-Richter and Mercalli scales, their hypocenters were almost at the same depth, and the damages they caused were almost at the same level.

The only difference between them was that the epicenter of the Northridge earthquake was in the middle of the densely populated central residence area of Reseda, while the epicenter of the Loma Prieta earthquake was in unpopulated area of a state park 127 km south from San Francisco Bay Area. The Northridge earthquake is the only large earthquake to originate directly under a major U.S. city in modern time.

Taking into account the location of the epicenter and the depth of the focus, it was normal to expect that the Northridge earthquake would be the most devastating ever earthquake to man-made constructions, or at least much more destructive than the Loma Prieta earthquake. How to explain the puzzling fact that these two earthquakes—one with epicenter 127 km away from the damaged area and the other one right in the middle of the damaged area—caused almost the same level of destructions? Were the seismic waves in the Northridge earthquake somehow attenuated, or were the seismic waves in the Loma Prieta earthquake somehow amplified?

There is a good reason to believe that San Fernando Valley was struck by body seismic waves coming directly from the 19 km deep hypocenter under Reseda. Simple geometrical calculation shows that San Francisco Bay Area could not be hit by bulk seismic waves coming directly from the 18 km deep hypocenter located under Nisene Marks State Park 127 km away from San Francisco Bay Area. It seems that during Loma Prieta and Northridge earthquakes different seismic phenomena caused the destructions.

The comparison of the parameters of these two earthquakes and their effects raises important questions.

First question: What is the physics of seismic waves? Classical seismology is based on propagation of elastic waves in elastic continuum and the general scattering theory (Aki and Richards 1980; Houdson 1980; Ben-Menahem and Singh 1981). The theory of continuum mechanics describes the deformations (strains) of objects and relate them to the resulting reaction forces (stresses) using linear algebra to perform coordinate constitutive modeling (mathematical description how the objects react to loadings based on 1st and 2nd laws of thermodynamics), calculate material derivatives, or determine principal stresses (Malvern 1969; Mase 1970; Landau and Lifshitz 1986). Most of the analyses in the classical continuum mechanics are based on linear elasticity theory which is limited to infinitesimal strains only while quadratic, cubic, and higher order strain terms are neglected. Applying linear elasticity theory to describe seismic wave propagation might be helpful to understand certain phenomena, but definitely it cannot be used to describe realistically an earthquake. Usually seismic wave have large amplitudes close to the focus, which make their propagation highly nonlinear. On the other hand they propagate in heterogeneous, nonlinear, dispersive media of the Earth's lithosphere and mantle. The recent development of advanced computer simulation and numerical solutions of nonlinear equations have prompted the development of nonlinear continuum mechanics and the introduction of new approach to the formulation of constitutive equations for nonlinear continua (Naugolnykh and Ostrovsky 1998; Dimitrienko 2011), however no coherent theory of nonlinear propagation of seismic wave has been developed yet.

Second question: How the surface seismic waves were able to keep their energy after traveling a long distance of 127 km and caused similar effect as the body waves that traveled only 19 km coming directly from the earthquake's focus in the Loma Prieta and Northridge earthquakes? Since surface seismic waves cannot be coming directly from the hypocenter, they are created and powered by body waves reaching the Earth's surface where they get converted into surface waves. This means that surface the seismic waves should carry lower energy than the body waves that power them. Comparing the parameters of Prieta Loma and Northridge earthquakes shows that this does not seem to be true. Was the attenuation on the Earth's surface very low or were the surface seismic waves somehow amplified due to nonlinear interactions?

Third question: What is the mechanism of generation of surface seismic waves from body waves that come to the Earth's surface? Why some body waves get converted into surface waves and others do not? It is well known that sidewinding surface waves known as Love waves are much more destructive than rolling surface waves known as Rayleigh waves? How Love and Rayleigh waves are created from body waves? Usually the tremor during an earthquake lasts only seconds. This hardly can be called a wave—a term that suggests a continuous, periodic, and lasting in time process. It would be more appropriate to consider this as a propagation of a pressure pulse or a series of pulses rather than a wave. Often the power of the pressure pulse is so high, especially close to the focus, that it has the characteristics of a shock wave. Because of the high nonlinearity of propagation wouldn't it reasonable to consider the existence of seismic solitary pulses or even seismic solitons?

Fourth question: Is there any relationship between geological structures in areas with seismic activities and propagation characteristics of body or surface seismic waves? If there is such a relationship, would it be possible to foresee the types of seismic waves and the amount of energy they carry if the geological topography of the area is known? Also, is the destruction of manmade constructions always due to the seismic power carried by oncoming seismic waves or certain geological characteristics may cause resonance phenomena, constructive interference, and local amplification of the waves? Typical example is the 1985 Mexico City earthquake where the lakebed sediments caused resonance effects from reflections from the edges of the basin making buildings from 6 to 15 stories in height to resonate in the frequency band of the lakebed motions (Murillo and Juan 1995). Many of these buildings had their upper floors collapsed leaving the lower floors undamaged.

Fifth question: What is a better strategy to protect cities from the destructive power of earthquakes? Is it building reinforced constructions able to survive the tremor or is it developing new techniques allowing oncoming seismic waves to be deflected away from the city or attenuated? Earthquake statistics show that many devastating earthquakes that had happened at various points of the Earth had not always been those of the greatest magnitude. Many earthquakes of moderate magnitudes had caused great amount of damage. It turns out that the magnitude of the earthquake is not the only factor causing the destruction. While the amplitude of the seismic waves certainly contributes to their destructiveness, the type of the seismic waves, the way they propagate, the direction of the displacement of ground masses and the nonlinear phenomena are of highest importance as well.

Today's strategy to fight destructive earthquakes consists mainly in designing earthquake resistant building constructions in hopes that they would be able to survive the seismic tremor. It is not possible to know in advance where the hypocenter of an earthquake would happen to be and what kind of seismic waves would be generated during the shock. Both technical difficulties and high cost make it impossible to design and build reinforced constructions able to face any type of seismic waves.

A different strategy to fight destructive earthquakes is discussed in this book, namely design of systems that redirect, attenuate, or scatter away oncoming seismic waves before they reach a city. Since surface seismic waves have proven to be much more destructive than bulk waves during earthquakes, the focus will be mainly on generation and propagation of surface seismic waves. Bulk seismic waves have been studied extensively in seismological research, while surface seismic waves have found less attention, especially the nonlinear wave propagation and nonlinear response of the rocks. Since surface seismic waves appear on the Earth's surface as a result of mode conversion of bulk seismic waves, we will limit our attention to bulk seismic waves as a source of surface seismic waves. Another reason to consider bulk seismic waves is that under certain conditions surface seismic waves can get converted back into bulk seismic waves redirecting in this way their energy flow down into the Earth's body and decreasing their destructive power on the surface. In this context we will discuss possibilities to create artificial systems for deflecting or attenuating surface seismic waves using methamaterials and other techniques.

Developing new earthquake-resistant designs and technologies for scattering, reflecting, or damping of surface seismic waves can prove to be both technically and financially more efficient way to fight the destructive power of earthquakes than relying solely on reinforced building constructions. Ideas of "cloaking" buildings from seismic waves (Farhat et al. 2009; Brun et al. 2009) have been launched as an analogy to aircraft cloaking against radars, i.e., hiding from electromagnetic waves of radar to avoid detection. Ideas to use seismic methamaterials to influence seismic waves passing through artificial anisotropic continuum have been explored (Kim and Das 2012). An experiment has been carried out in France near Grenoble (Brûlé et al. 2014). Since the soil is a heterogeneous, nonlinear, dispersive medium, offering various uncertainties, the objective of the Grenoble experiment was to find realistic input values for the simulations by adjusting the soil parameters, such as shear modulus, Q-factor, etc.

Despite the encouraging start, it seems that the task of designing universal seismic-wave deflecting or seismic-energy absorbing systems able to protect cities from any type of seismic waves coming from any direction is a quite difficult task. The Earth's 100 km tick lithosphere where most seismic activities take place is a very heterogeneous and nonlinear medium in which high-amplitude nonlinear seismic waves propagate. The rocks that take part of the composition of the lithosphere have a very diverse structure and exhibit strong nonlinearity. Linear wave propagation in homogeneous elastic continuum is well understood. Nonlinear wave propagation in homogeneous elastic materials is a complex phenomenon which has been extensively researched, cumbersome mathematical model have been developed, and most of the problems have been solved only numerically using various simplifying approximations. Nonlinear wave propagation in dispersive nonlinear media remains a problem of extreme complexity. Seismic wave propagation at long distances causing earthquakes is possible only because of nonlinear phenomena such as generation of higher harmonics and subharmonics, self-modulation, parametric amplification, wave-wave interactions, self-focusing, etc. Earthquakes with their typical characteristics as we know them can occur in a linear world.

The goal of this book is to explore the possibilities to use acoustoelectronics and acoustooptics experimental methods and techniques for designing systems that are capable to protect cities by not allowing any type of seismic waves to reach them. The seismic waves are analyzed from the point of view of the nonlinear wave propagation theory and physical acoustics—a branch of solid state physics that studies the properties of elastic waves in solids and fluids as a result of their microscopic structure. Solid state physics uses quantum mechanics tools to study large-scale electromagnetic, optic, mechanic, crystallographic, and metallurgic properties of condensed matter in connection with its atomic-scale properties. Physical acoustics techniques are used for nondestructive testing (NDT) of materials, medical ultrasonic imaging, signal processing, ocean acoustics, acoustoelectronics (delay lines, convolvers, filters), and acoustooptics (light modulators, laser deflectors, and Q-switches of laser beams).

Physical acoustics is a debtor to seismology. Surface acoustic waves (SAW) used in many acoustoelectronics devices have been discovered by scientists in the process of studying seismic waves. Rayleigh studying seismic waves published his "On Waves Propagated along the Plane Surface of an Elastic Solid" (Ryleigh 1885). Seismologists at that time were skeptical about Rayleigh's findings until 1911 when Love published "Some Problems of Geodynamics" (Love 1911) where he reported a mathematical model of surface seismic waves known as Love waves. The British seismologist Robert Stoneley discovered the surface interface waves Stonley waves in 1924 (Stoneley 1924).

The piezoelectric effect was discovered in 1880 by French physicists (Jacques and Pierre Curie 1880). Since then many works have been published about bulk and surface elastic waves in piezoelectric and non-piezoelectric solids laying out the base of physical acoustics, acoustoelectronics, and acoustooptics (Viktorov 1967; Farnell 1970; Farnell and Adler 1972; Auld 1973; Dieulesaint and Royer 1974; Biryukov et al. 1995; V'yun and Yakovkin 1990) in parallel with the development of the theory of elasticity (Landau and Lifshitz 1986). Theory of linear elasticity has been used in many works to describe seismic wave propagation (Aki et al. 2002; Ben-Menahem and Singh 1981; Chapman 2004), but physical acoustics has found few applications in seismology. Maybe now it is time physical acoustics—the science of elastic waves in solids to return the favor to seismology—the science of seismic waves in the Earth by implementing some of the theoretical and experimental methods and techniques that have been used for designing acoustoelectronics devices into developing earthquake-resistant design and technology.

Linearity is only an approximation of nonlinearity used to simplify the extreme complexity of the nonlinear systems and get exact solutions of the governing differential equations. However, many important phenomena of the real world can be easily overlooked by using solely the linear theory. Such phenomena are 'wave generation by wave', self-modulation, resonance of high-order harmonics, self-focusing, or parametric amplification of high-amplitude elastic waves propagating in dispersive nonlinear media (Mayer 1995; Naugolnykh and Ostrovsky 1998; Zabolotskaya 1992; Maradudin 1990; Lomonosov 1999). Laser techniques enabled generation of very high amplitude pulses with acoustic Mach numbers about 0.01 (Kolomenskii et al. 1997a, 2003b). Such waves drive the medium into the nonlinear elastic regime and shock fronts can be formed during their propagation. As an intense surface acoustic wave (SAW) propagates, the temporal evolution of the wave shape provides information on the nonlinear acoustic parameters and the nonlinear elastic constants of the material.

Physical acoustics usually deals with both linear and nonlinear elastic wave propagation in homogeneous isotropic materials (fused quartz, glass, etc.) homogeneous anisotropic materials such as single crystals (α-quartz (α-SiO$_2$), sapphire (α-Al$_2$O$_3$), lithium niobate (LiNbO$_3$)), and heterogeneous and nonlinear materials (composites, ceramics, defected solids, etc.). The generation, propagation, and detection of linear and nonlinear elastic waves in various linear and nonlinear materials have been studied extensively in order to build acoustoelectronics and acoustooptics devices with wide range of applications in nondestructive testing of materials, signal processing, delay lines, mechanical filters, piezoelectric resonators for frequency control, laser-beam modulators, diffraction gratings, actuators, chemical and bio-molecular sensors. Acoustoelectronics nonlinearities of surface acoustic waves resulting from couplings of displacement fields (V'yun and Yakovkin 1990) and propagation in heterogeneous media (Biryukov et al. 1995) have attracted great interest. In the process of studying, designing, manufacturing, and testing systems for nondestructive testing of materials and medical diagnostics significant knowledge has been accumulated about the physical properties of various types of elastic waves and their propagation in isotropic, anisotropic, and heterogeneous materials. This knowledge can be used for better understanding of nonlinear seismic systems and developing experimental modelling solutions that can lead to practical applications.

1

Origin of Earthquakes and Seismic Waves

During a disturbance deep in the Earth's crust or mantle such as faults, tectonic plate shifts, volcano eruptions or underground explosions, a large amount of energy is released as a result of which seismic longitudinal (P-waves), shear (S-waves), and surface waves propagate, causing an earthquake. This cause-and-effect process seems simple and easy to understand. However for describing step by step the generation and propagation of various types of seismic waves, many factors should be taken into consideration. Deep seismic disturbances cannot be directly studied, but observations of man-caused explosions causing tremors point to the generation of short, powerful, pressure pulses reaching Mega or even Giga Pascal ranges in the focus of the earthquake. However, a pressure pulse is not a wave. A wave is a periodic transient motion in a medium of propagation that requires a continuous driving force. The question is how a pressure pulse becomes the origin of a seismic wave? Of course, in an earthquake's focus not one, but a raw of pressure pulses can occur. They follow each other at random intervals of time and the question of the origin of seismic waves is still relevant.

In this chapter we will discuss the mechanism of the generation of various kinds of transient seismic waves from a high energy pressure pulse that are able to travel long distances in a heterogeneous, dispersive, and dissipative media causing what we call an earthquake.

1.1 Statement of the problem

Seismic waves are usually assumed to be elastic waves that propagate in the Earth's body. When an elastic wave propagates in some material it applies periodic mechanical force per unit of surface $\vec{F}_1$ (stress) resulting in a periodic deformation (strain). The material reacts to the deformation with a restoring force $-\vec{F}_2$ trying to bring the material's initial shape back to its equilibrium state. The theory of linear elasticity is restricted to infinitely small deformations which disappear when the causing force $\vec{F}_1$ stops acting and the restoring force $-\vec{F}_2$ (where $|\vec{F}_1| = |\vec{F}_2|$) brings the elastic material back to its initial sate of equilibrium following a linear stress-strain relationship. An elastic wave is considered to be linear if its amplitude is infinitesimal and the nonlinear terms of the amplitude in the governing equation describing the propagation of the wave can be neglected, i.e., linearizing in this way the nonlinear differential equation of motion. In an infinite linear medium of propagation a plane linear elastic wave keeps its waveform constant during the propagation. The change of the internal energy (kinetic, potential, thermal, etc.) is proportional to the deformation that the wave causes, i.e., the change of internal energy density is a linear function of the deformation.

If the elastic wave amplitude becomes larger, the linear approximation does not hold anymore and the wave propagation becomes nonlinear. A nonlinear finite-amplitude elastic wave propagates in a very different way than a linear small-amplitude elastic wave. Various portions of the wave travel at different velocities causing distortions of the waveform. If the medium of propagation is nonlinear, it can affect additionally the wave's characteristics because of scattering from heterogeneities, dispersion, or dissipation. Many factors can cause dissipation of elastic energy—internal friction, thermal phenomena, absorption, scattering, etc. Dispersion is a phenomenon when the phase velocity of the wave depends on its frequency. Dispersion can be due either to specific geometry features in the medium of propagation—waveguide structures, boundaries, or layers, or to its heterogeneous structure, scattering characteristics, or dissipation of elastic energy. Introducing nonlinear wave propagation instead of linear approximation makes the problem much more complicated but also more realistic.

When finite-amplitude elastic wave propagates in homogeneous, nondispersive, dissipative media, it is sufficient to use quadratic nonlinearity—keeping the second-order terms related to the wave amplitude (deformation) and ignoring the third, fourth and higher order terms. However, introducing heterogeneities in the medium of propagation such as grain structures, cavities, and cracks increases significantly the challenge and the problem cannot be reduced anymore only to quadratic approximation (Naugolnykh and Ostrovsky 1998). In a dissipative and dispersive medium of propagation, a finite-amplitude elastic wave cannot keep its waveform unchanged during the propagation. Its shape gets distorted due to the differences in the velocities of various parts of the wave (Fig. 1.1.1). The nonlinearity produces harmonic components which interact with each other by exchanging energy and giving rise to more harmonics and waves of combination frequencies. If

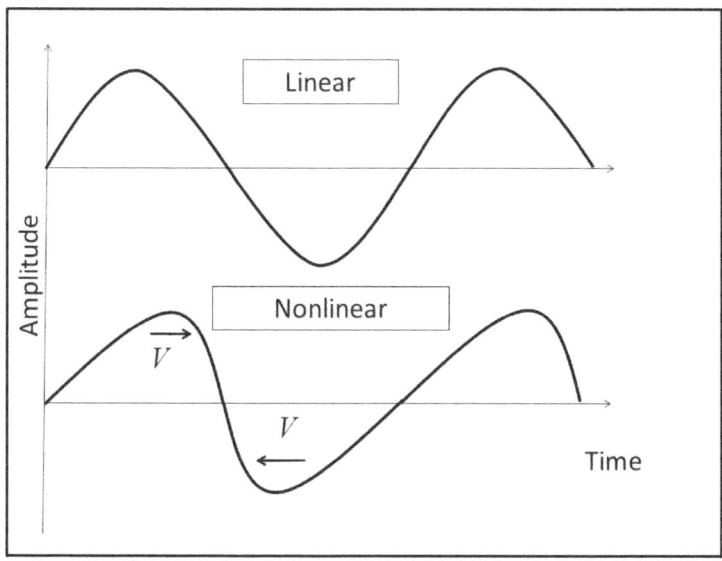

Fig. 1.1.1. Distortion of the profile of a high-amplitude elastic wave as it propagates.

the nonlinearity is weak the medium responds by generating new waves at combination frequencies that propagate, i.e., when the synchronism conditions are satisfied, a space-time resonance occurs. In the absence of dispersion the synchronism conditions are satisfied at even great number of frequencies and even more harmonics grow.

In the presence of dispersion, low harmonics grow because they satisfy the linear dispersion relation and cause waveform steepening. High harmonics are suppressed because they deviate from the linear dispersion relation. A stable solution is obtained by balancing the waveform steepening due to nonlinearity and the dispersion effect of higher order terms where the dispersion relation is not linear.

Materials subjected to large deformations behave in a nonlinear way (Fig. 1.1.2). The nonlinearity in such cases is due either to the elastic properties of the material or to the large forces causing the waveform deformation, as for example propagation of high-amplitude elastic waves. The wave front of a small-amplitude elastic pulse wave in heterogeneous medium of propagation gets quickly deformed and the wave loses its energy similarly to as it were propagating in a high-attenuation dissipative medium. However, elastic heterogeneous media such as rocks can support steady high-amplitude elastic waves due to wave dispersion caused by scattering with the heterogeneous structure (Grady 1997). Rocks in the Earth's lithosphere, where most of the seismic activities take place, are heterogeneous with a highly nonlinear behavior due to various chemical compositions and structures. The nonlinear

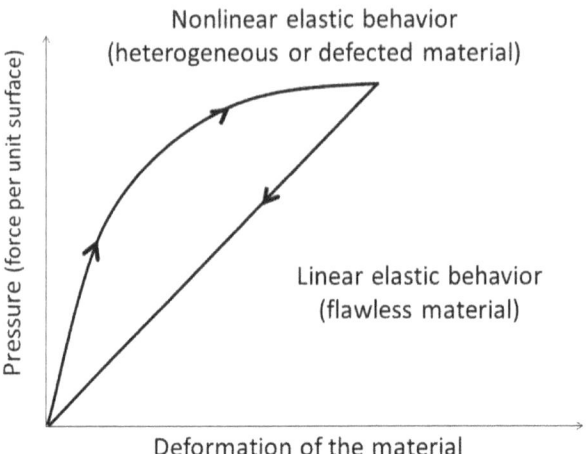

Fig. 1.1.2. Linear and nonlinear elastic wave propagation.

behavior of rocks is caused by intergranular cracks, dislocations, and weak or failing grain contacts in poorly consolidated sediments (Fjær et al. 1992). Typically, rocks show nonlinear behavior for strain amplitudes exceeding approximately 10^{-6}. The strain amplitudes of elastic waves in artificial seismic explorations and elastic pulse transmission experiments are usually small. However, sedimentary rocks may have large nonlinear elastic coefficients compared to well-ordered crystalline solids, such that nonlinear effects may be observable and important. Because of cracks and poor consolidation, rocks may have large third- and fourth-order nonlinear elastic moduli. For example, third-order nonlinear compressional elastic coefficient in Berea Sandstone—a sedimentary rock whose grains are predominately sand-size and are composed of quartz held together by silica—has been found to be three orders of magnitude larger than the linear coefficient (Sinha and Winkler 1999) resulting in strong nonlinear elastic wave propagation (Meegan et al. 1993). The high porosity and permeability of Berea Sandstone rock makes it a good reservoir of fluids (oil or water). Also its effective Young's modulus has been found to depend strongly on the elastic wave amplitude and frequency (Tutuncu et al. 1998a,b). The nonlinear wave propagation equation must be solved numerically, even in a homogeneous medium if constitutive relation contains third- and fourth-order nonlinear terms in addition to the linear term (Hokstad 2004).

Finite-amplitude elastic waves that propagate in a nonlinear, dispersive, and heterogeneous medium can change the medium's elastic properties and vice versa—the medium can change the wave's parameters as well. Various phenomena occur in these conditions that otherwise do not exist in the linear approximation such as scattering of elastic wave from other elastic wave, self-modulation, parametric wave amplification, self-focusing, elastic beam bonding, wave trapping, etc.

Classical seismology is based on linear elastic wave propagation in linear, homogeneous, elastic, nondispersive, isotropic and anisotropic media as defined by the continuum mechanics theory and the general wave scattering theory using Green function method, so most of the phenomena related to the nonlinear elastic wave propagation as well as to the heterogeneous, nonlinear, dispersive, and dissipative properties of the medium of propagation are left unaccounted for. Seismic wave propagation is possible only because of nonlinearity. A linear nondispersive world is earthquake-free. An earthquake as we know it cannot originate from linear elastic waves that propagate in a linear, homogeneous, nondispersive medium.

1.1.1 What causes an earthquake?

Nonlinearity of seismic waves and rocks has not been fully understood mainly because the nonlinear wave processes in other scientific fields such as acoustics, optics, electromagnetism, atomic, and nuclear physics are in process of development. It would be impossible to develop reliable earthquake-resistant design and technology without studying seismic waves as nonlinear elastic waves propagating in nonlinear rock media.

In this section we will show that various wave processes have similar behavior and, therefore, can be studied using equivalent circuits. Mechanical systems can be described using electrical equivalent circuits, optical equivalent circuits, or seismic equivalent circuits. This is possible because various wave phenomena have the same basic characteristics—wavelength, frequency, velocity of propagation, etc. and can interact with each other in resonance conditions. The only difference between various waves is that they transport various types of energy—elastic, electric, magnetic, optic, thermal, etc.

As it was mentioned already an earthquake is caused by a sudden disturbance in the Earth's body during which a large amount of energy is released and a pressure pulse is generated. Note that instead of using the word 'wave' we emphasize on the word 'pressure pulse'. The reason is, as it has been mentioned already, that a pressure pulse is not a wave. A wave is a periodic motion caused by a period driving force with specific parameters such as frequency, velocity of propagation, and wavelength bound by the relation $v = V/\lambda$. The pressure pulse has neither frequency, nor wavelength (Fig. 1.1.1.1). It travels with the velocity of sound of the material in which it propagates, or faster if it is a shock pulse. An earthquake can be linked to effects similar to when an anvil is hit with a hammer. After the hammer makes contact with the anvil a primary pressure pulse will start moving into the anvil inwards from the point of contact. The anvil's surface will bounce up into negative pressure

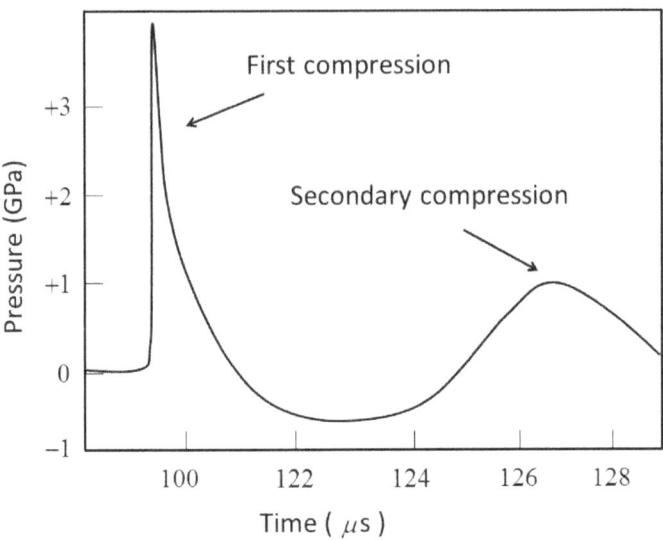

Fig. 1.1.1.1. Pressure pulse waveform at the focus.

followed by a secondary compression with positive pressure that will vanish quickly without any wave being generated if the anvil is large enough. More visually this process can be linked to the effect of a stone thrown in water. When the stone hits the water surface a row of ripples start moving outwards of the point of contact. The ripples are caused by the water initially going up around the hole made by the stone when it hits the surface and then going down to fill the hole left by the stone. This up-and-down water motion is repeated a couple of times with decreasing amplitude and finally vanishes because of the water's viscosity. A water wave packet moves outwards from the center of disturbance. A wave packet can be presented as a Fourier series—an infinite set of harmonic (for example sinusoidal) waves of different frequencies (or wavenumbers, or wavelengths) with phases such that the waves interfere constructively in a narrow space and destructively everywhere else. In quantum mechanics a wave packet is associated to a particle. According to the Heisenberg uncertainty principle the position and momentum of a particle cannot be known simultaneously. This means that a specific wave with a fixed wavelength cannot represent the particle. Instead a superposition of waves with different wavelengths ranging closely to the central wavelength value of a wave packet is associated to the particle. The wave packet is the wave function of the particle. Each constituent wave (and also the wave packet) is a solution of the wave equation of motion (Schrödinger's equation). Wave packets having Gaussian shape have been used to analyze water waves (Mei 1989). Wave packets and their Fourier components are always solutions to the corresponding equation of motion. As it will be shown in Section 1.1.3 the

Fourier transform of a Gaussian function is a Gaussian function itself. If the equation of motion describes nondispersive wave propagation the waveform of the wave packet remains constant during the propagation. In presence of dispersion the waveform of the wave packet changes as the packet propagates.

While moving with constant velocity in a dispersive medium the wave packet is delocalizing rapidly, its width is increasing with time, so eventually it diffuses to an unlimited region of space. The wave packet does not have a driving force necessary to maintain its propagation as a continuous wave in the medium of propagation. So, how the seismic waves are generated and where are they coming from?

We presented the two cases where pressure creates wave packets in an anvil hit by a hammer and in water by a stone dropped on the water surface, but these two cases are very different. The difference is not because the anvil is a solid whereas water is a fluid, but because standing waves can be created in the anvil from the energy of the pressure pulse, while the wave packet in the water will vanish in a short distance from the point where the stone fell onto the water surface. The pressure pulse in the anvil undergoes multiple reflections from the anvil's side walls transferring its energy to resonance standing waves that vibrate at the anvil's normal frequencies and their harmonics in a frequency spectrum defined by the shape, dimensions, and metal of the anvil. The live span of these standing waves is defined by the absorption properties of the anvil's metal. A standing wave vibration is not a transient wave. After some time dissipation phenomena will attenuate the standing waves. However, if the anvil is in contact with other low-loss materials the elastic energy of the standing waves will be transferred to them and they will start vibrating at their own normal frequencies.

It is well known that two types of seismic waves propagate through the Earth's body—longitudinal bulk (or body) waves called also P-waves and shear (or transverse) body waves called also S-waves. Also on the Earth's surface propagate various types of surface elastic waves displacing horizontally or vertically the ground. The P-wave is usually called primary wave because it is the fastest wave that supposedly arrives first at the seismographic station. The S-wave is called secondary wave because it is slower than the P-wave and it arrives after it. These waves are considered to be direct waves arriving from the focus of the earthquake. The last wave arriving at the seismic station is considered to be a surface seismic wave which is slower than the S-wave. In Fig. 1.1.1.2 a seismogram shows the vertical earth displacement with the supposed locations of the P- and S-waves indicated with arrows assuming first and second arrivals. Seismograms of vertical displacement of the ground are not very informative about the type of seismic waves. More information provide seismograms of vertical, radial, and transverse components of the displacement because they help to identify the wave polarization (direction of displacement). The three seismograms in Fig. 1.1.1.3 (Oklahoma Geological

BKS - Vertical Ground Movement
Ground Displacement [mm]
0.12
0.1
0.08
0.06
0.04
0.02
0
−0.02
−0.04
Δt = 3.5 s
P
S
04:00:15
04:00:20
04:00:25
04:00:30
Time

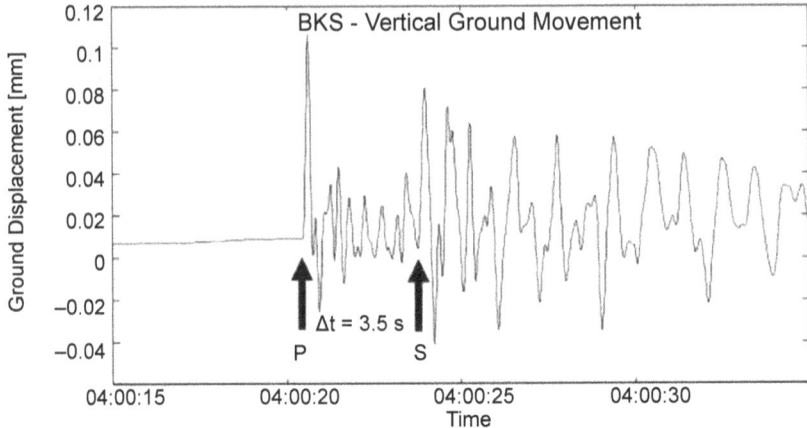

Fig. 1.1.1.2. A seismogram showing the vertical earth displacement with the supposed location of the P- and S-waves.

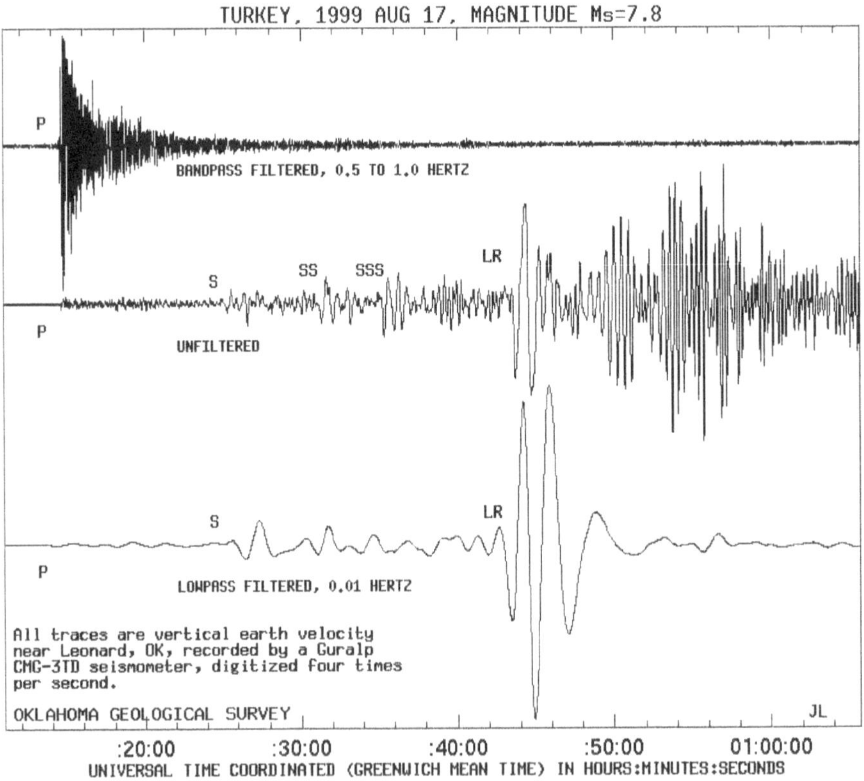

TURKEY, 1999 AUG 17, MAGNITUDE Ms=7.8
P
BANDPASS FILTERED, 0.5 TO 1.0 HERTZ
S
SS
SSS
LR
P
UNFILTERED
S
LR
P
LOWPASS FILTERED, 0.01 HERTZ
All traces are vertical earth velocity near Leonard, OK, recorded by a Guralp CMG-3TD seismometer, digitized four times per second.
OKLAHOMA GEOLOGICAL SURVEY
JL
:20:00
:30:00
:40:00
:50:00
01:00:00
UNIVERSAL TIME COORDINATED (GREENWICH MEAN TIME) IN HOURS:MINUTES:SECONDS

Fig. 1.1.1.3. Three seismograms showing the vertical earth velocity near Leonard Oklahoma caused by the August 17, 1999 earthquake in Turkey (Oklahoma Geological Survey Earthquake Catalog).

Survey Earthquake Catalog) show the vertical earth velocity near Leonard Oklahoma. The middle trace is the unfiltered output of the seismometer. It shows P, S and LR (LR are surface Rayleigh waves). It also shows SS-waves (reflected from the earths surface midway between the epicenter and Leonard) and SSS-waves (reflected twice from the surface). The top trace is the same seismogram filtered to pass only waves with frequencies between 0.5 and 2.0 Hertz. This shows only the P seismic waves. The bottom trace is the same as the middle trace except that it is filtered to only pass waves with frequencies below 0.01 Hertz. These extremely low frequency waves vibrate the ground so slowly that the time from one peak to another peak is 100 seconds or more. The seismograms in Fig. 1.1.1.4 show ground velocity recorded in vertical, radial (toward the epicenter), and transverse (at right angles to the direction of the

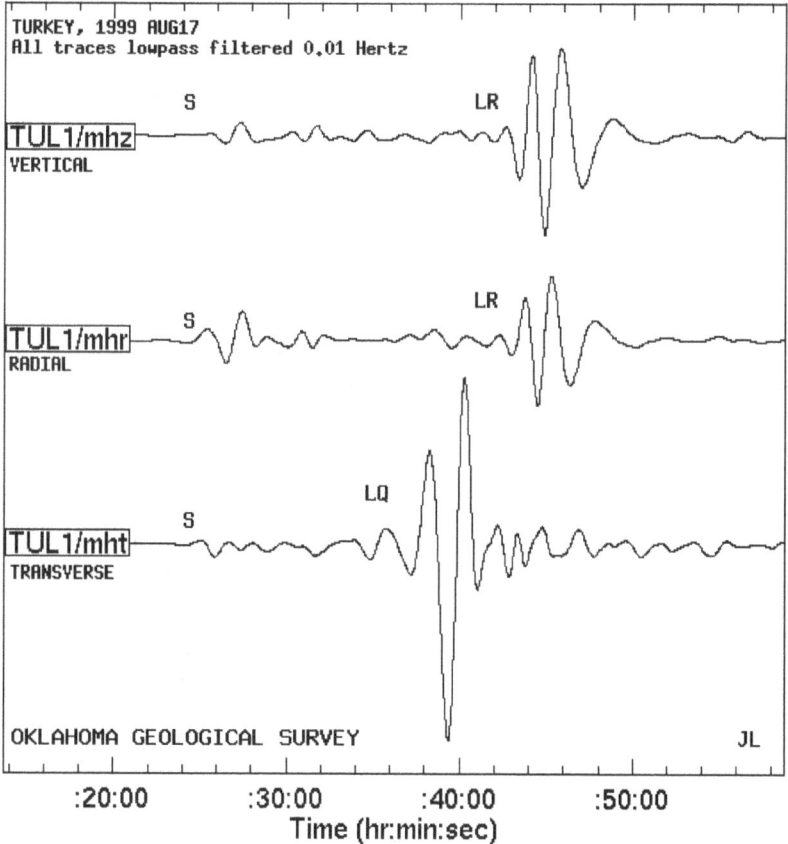

Fig. 1.1.1.4. The seismograms show ground velocity recorded in vertical, radial (toward the epicenter), and transverse (at right angles to the direction of the epicenter) directions registered during the August 17, 1999 earthquake in Turkey (Oklahoma Geological Survey Earthquake Catalog).

epicenter) directions. Each trace is low-pass filtered to show only waves with a period of 100 seconds or more. The P arrival is off the figure. S is clear on all traces. When the traces are rotated to produce vertical, radial, and transverse motion, the two types of surface waves are completely separated. Rayleigh waves (LR) show only on the vertical and radial traces. The earlier arriving Love (LQ) waves only show on the transverse seismogram. The vertical LR ground motion was about 3.8 millimeters peak-to-peak, at the underground vault near Leonard. This allowed the OGS to calculate an MW type magnitude of 7.8. It might seem that if the earth's surface in Oklahoma moved up and down four millimeters, that buildings would be damaged, but they were not. OGS explained that there would be damage if the movement were rapid. However, the movement was like slowly lifting the earth over a minute or so, then lowering it as slowly. Such slow motion will not effect anything but the detection sensor of a very broadband seismometer.

Seismic waves identification by the time of their arrival seems to be an easy and convenient way to interpret seismograms. However, seismic wave propagation is a complex process and in many cases such identification could be inaccurate. Real seismic waves are never pure P-type or pure S-type, they are quasi-longitudinal (qP) and quasi-shear (qS). They propagate independently but their phase velocities and group velocities are not collinear. The two velocities make an angle between them which means that the wave propagates in certain direction and its energy is transported in a different direction. Since nonlinearity and dispersion strongly modify the seismic wave propagation, the identification of various types of seismic waves and the interpretation of the seismograms need to be adjusted accordingly. Later in the book we will come back to the seismograms in Figs. 1.1.1.3 and 1.1.1.4 when we be discussing nonlinear body and surface wave propagation. Beside these three types of waves there is another type of waves called coda waves, or just coda, seen at the tail of the S-wave on the seismogram in Fig. 1.1.1.5. It is interesting to notice that in Fig. 1.1.1.5 not only the tail, but also the whole wave packet is full of coda. Linear theory cannot explain coda waves. As we will see nonlinear theory provides a simple and plausible explanation of these waves.

The seismic coda waves are believed to result from backscattering processes from numerous heterogeneities distributed uniformly in the Earth's crust (Aki and Chouet 1975). Looking at the seismograms shown in Figs. 1.1.1.2, 1.1.1.3 and 1.1.1.4 two questions arise: 1) if coda waves were backscattered waves why the ballistic S-wave is having coda not only in its tail but also in its body or wave packets with same for the P-wave, and 2) since coda-like waves fill the wave packets aren't coda waves forming the wave packets of the ballistic waves and moving together with them? These two questions about the origin of coda raise also a third question. Can the waves reaching a seismic station can be identified unambiguously as P-, S-, or L-waves according to the time of their arrival or their 'coda' filling should also be taken into consideration?

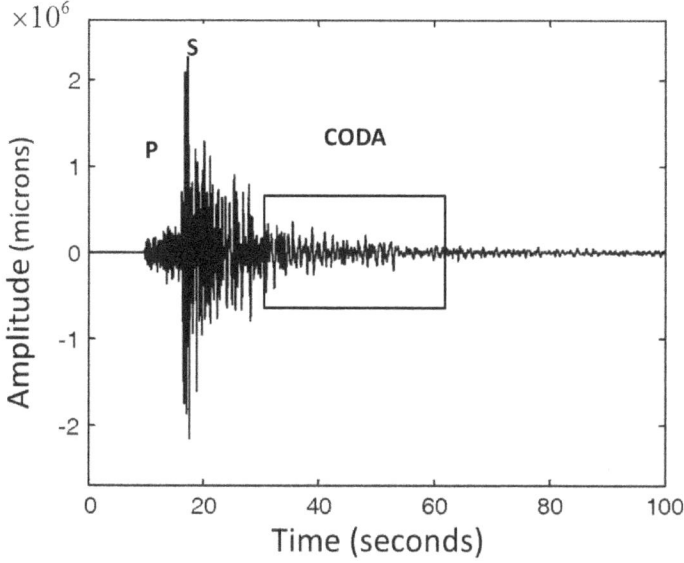

Fig. 1.1.1.5. Coda waves.

These three questions are relevant because it seems that 'coda waves' can be P-, S-, or LR-waves with different periods as it can be seen from the seismogram in Fig. 1.1.1.3.

It is known that the seismic waves' frequencies determine how far the wave can travel and how damaging they are to man-made structures. The spectrum can go from 0.001 Hz to several hundred Hz. Low-frequency waves can travel longer distances, but they usually cause less damage than the high-frequency waves that tend to dissipate fast. The most damage is caused by waves in the frequency band between 1 Hz and 10 Hz which can still travel long distances before get dissipated. So, how the frequency bands of seismic waves are defined if all of them originate from the energy of a pressure pulse in the earthquake's focus? When we hit a tone on the piano board the small hammer under the hood hits the corresponding string generating a single pulse that transfers its energy to the string and force it to vibrate at the tone frequency. The string is attached firmly at its ends to the piano cast iron plate, so it vibrates as a standing elastic wave at its fundamental normal frequency defined by the length of the string and the velocity of the wave in the string material. The elastic energy of the vibrating string is transmitted from the cast iron plate to the piano wood soundboard and into the air as a longitudinal (there are no shear waves in fluids) sound wave that we hear as a single tone. The frequency of the sound wave that we hear is the same as the frequency of the vibrating string but its wavelength and velocity of propagation are different in the air.

The mechanism of elastic waves generation when the hammer hits the metal anvil is the same as in the case of the piano. However, while the piano produces pure tone sound waves at well-defined frequencies, the anvil produces noise—a set of normal frequencies of the anvil. We can make a conclusion that the standing elastic wave frequency is defined by the resonator—either the anvil or the piano—but not by the hammer.

Now we need to clarify how the standing wave vibrations are transformed into transient waves that propagate in the air and deliver music (or noise in the case of our anvil) to our ear. To generate an elastic wave a permanent driving periodic force is needed. For example in nondestructive testing or medical ultrasound imaging piezoelectric transducers are used for the generation of ultrasonic waves. No such force exists in the case of an earthquake. A short pressure pulse (or pulse train) is produced at the fault and then the process stops. In a simple linear system the spectrum of a pulse is its Fourier transform. The spectrum of the pulse in Fig. 1.1.3.1 is shown in Fig. 1.1.3.2 in the Section 1.1.3. The pressure pulse will undergo numerous reflections on the boundaries with adjacent rocks generating local vibrations in the form of standing waves but apparently no transient waves will be created. Indeed, this could happen only if the rock were isolated—for example in a vacuum chamber. In this case the energy released at the fault would remain trapped in that rock without spreading any farther. However, the rock is not in vacuum; it is immerged in a medium of adjacent rocks with different elastic properties separated from each other by clay or sandstone forming a complex system of composite resonators. Standing wave vibrations will be transferred from rock to rock creating new sets of resonance frequencies in a wide spectrum. Let us consider a simple composite resonator formed by two adjacent media as shown in Fig. 1.1.1.6 (Miller and Bolef 1968; Lu 1974). Let assume that only one P-wave is traveling between the two blocks with velocities V_1 and V_2, fundamental mode resonance frequencies ω_1 and ω_2, and elastic impedances $Z_1 = \rho_1 V_1$ and $Z_2 = \rho_2 V_2$, respectively (ρ is the material density). If one assumes a total reflection of waves at both end surfaces, a reflection coefficient $r = \dfrac{Z_1 - Z_2}{Z_1 + Z_2}$ between the two media, and neglects the acoustic losses in both media 1 and 2, the fundamental resonance frequency of the composite resonator v_c can be determined by solving the following transcendent equation:

$$
\begin{aligned}
&2r\left[\cos\left(2\pi v_c / v_1\right) - \cos\left(2\pi v_c / v_2\right)\right] + \\
&\left(1 + r^2\right)\left[1 - \cos\left(2\pi v_c / v_2\right) \cdot \cos\left(2\pi v_c / v_1\right)\right] + \\
&\left(1 - r^2\right)\sin\left(2\pi v_c / v_2\right) \cdot \sin\left(2\pi v_c / v_1\right) = 0
\end{aligned}
\qquad (1.1.1.1)
$$

From Eq. 1.1.1.1 we can see that a composite resonator system resonates at different resonance frequencies than its components. We can imagine the complexity of the problem if we had a composite resonator built from a great

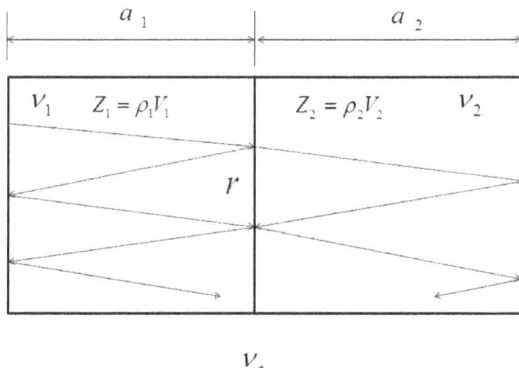

Fig. 1.1.1.6. Composite resonator.

number of individual resonators each one resonating at its own resonance frequencies and contributing to the formation of the composite resonator spectrum.

The diffuse elastic field generated by a large number of individual rock resonators has chaotic characteristics. The Earth's body and surface are in a state of a permanent vibration. During earthquakes the amplitudes of the vibrations increase. Nonlinearity and dispersion of the rock contribute to seismic beat and propagation of low-frequency wave packets that modulate high-frequency carrier waves of the diffuse elastic field. This diffuse elastic field can be defined as 'coda'—the building component of the wave packets, not just their tails. The term 'coda' is probably not even appropriate to be used anymore in this aspect because, as shown in Fig. 1.1.1.5, traditionally it is used to characterize the tail of a body wave. Despite of these considerations will continue using the term 'coda wave' or just 'coda' but instead of thinking of it as a 'tail wave' we will define it as a diffuse elastic field composed of body or surface elastic waves of various types forming a set of carrier waves. Actually it is interesting to note that 'coda' (Italian for 'tail', plural code) is a term used in music to designate a passage that brings a piece to an end. In terms of the formation of a seismic diffuse elastic field instead of being a tail, coda is a set of carrier seismic waves transporting most of the seismic energy. Coda gets bandpass filtered through large number of scattering processes and amplified by various nonlinear mechanisms. Its average amplitude increases during an earthquake not only because of influx of energy but also because of various nonlinear and dispersion mechanisms of amplification. Nonlinearity and dispersion cause the generation seismic waves of combination frequencies leading to seismic beat and self-modulation. Low-frequency P-, S-, and surface waves modulate by amplitude, frequency, and phase high-frequency coda waves. This self-modulation process creates wave packets formed by wave multiplication of carrier waves and modulating signals that are registered on the seismograms. Body coda and surface coda have different origins and different characteristics. Due to heterogeneity of the rock and variable levels of nonlinearity and dispersion, the modulation of the carrier waves by modulating body and surface waves can occur at various points located at various distances from the seismic stations where the seismograms are registered. This means that a modulating S-wave created nearby a seismic station can arrive before a modulating P-wave that has been created at a different point located at a farther distance from the seismic station. Such an S-wave can also be interpreted as an SS or SSS wave as well. This explains the differences observed in seismograms registered during the same earthquake at different seismic stations. Figure 1.1.1.7 is presenting seismograms registered by different seismic stations during the 1999 Kansas City collapse event. Three seismograms of vertical earth motion were recorded near Leonard, OK (above), near Vivian, OK (middle), and near Slick, OK (low). The three seismographs are 368 km/229 miles (Leonard), 400 km/249 miles (Slick), and 429 km/267 miles (Vivian) from the event. It is clear from these three seismograms that it is hard to identify unambiguously P-, S-, or LR-waves according to the arrival times. However, clearly the frequencies of coda waves are different. Since P-, S-, and LR-waves are independently propagating we assume that a specific type of carrier waves can be modulated by the same type of modulating wave only. This means P-carrier waves can be modulated by P-waves only, S-carrier waves by modulating S-waves only,

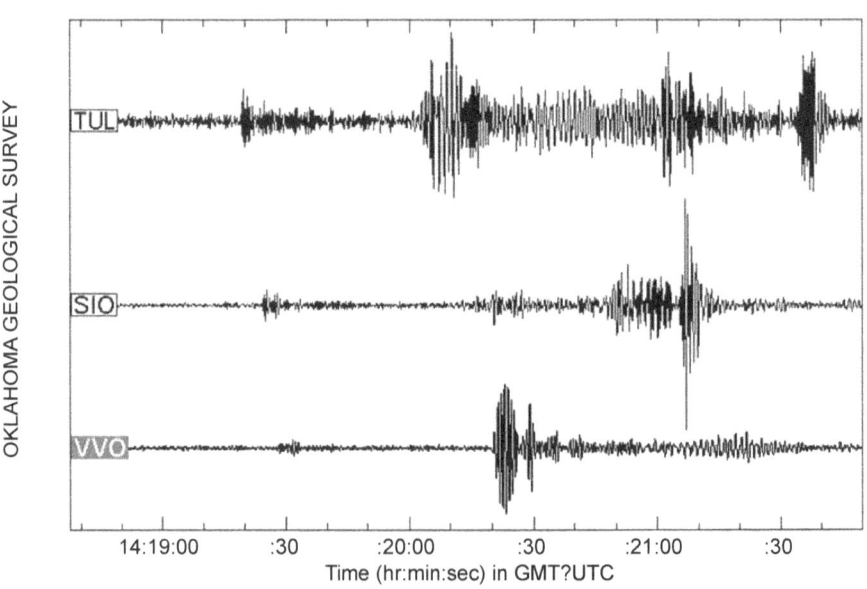

Fig. 1.1.1.7. A seismogram from the 1999 Kansas City collapse event.

and surface carrier waves can be modulated by a modulating surface waves only. Figure 1.1.1.8 shows seismograms of the January 10, 2010, earthquake in Haiti registered by a Manitoba (Canada) station. Radial, transverse, and vertical ground displacements shown on these seismograms can be interpreted as caused by seismic waves modulating coda waves.

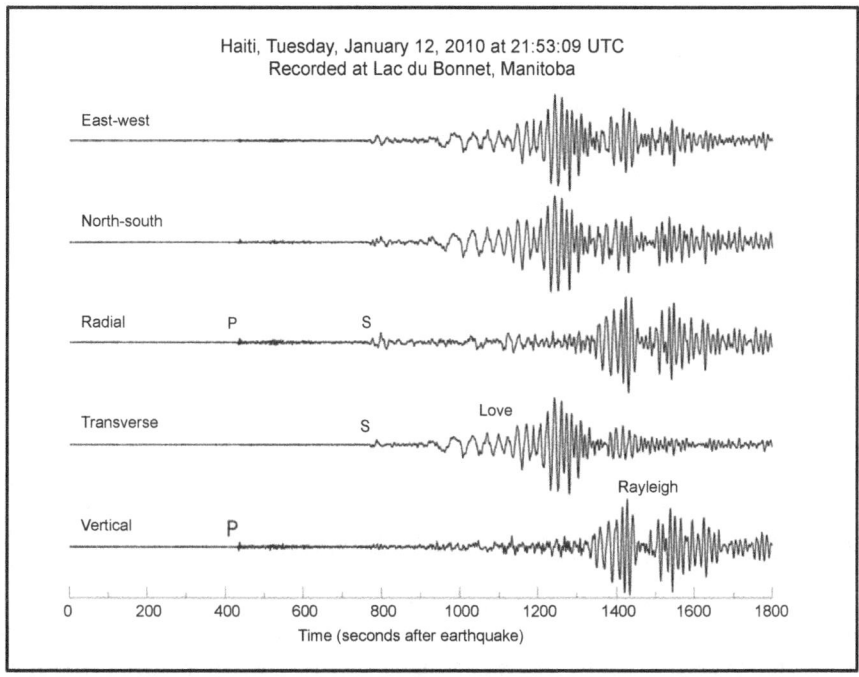

Fig. 1.1.1.8. Seismograms of the January 10, 2010, earthquake in Haiti registered by a Manitoba (Canada) station. Radial, transverse, and vertical ground displacements shown on these seismograms can be interpreted as caused by seismic waves modulating coda waves.

The processes during the earthquake are shown in Fig. 1.1.1.9. Let's suppose that the earthquake's 'hammer' gives a blow in the focus of the earthquake creating a high energy pressure pulse. Of course, there is no piano down there, but something that is similar to the anvil—randomly shaped rocks that resonate at their normal frequencies in a wide frequency spectrum forming a diffuse elastic field or ground noise. This diffuse noise field is not transient and, therefore, it is not coda. It is an elastic field created by a large set of standing waves. Coda is transient waves that propagate in the medium and will be subjected to intensive scattering, attenuation and filtering by the heterogeneous rock and not all waves that compose the elastic field will be able to propagate far from the focus of the earthquake. Weak waves and high-frequency waves will decay through scattering and absorption, but

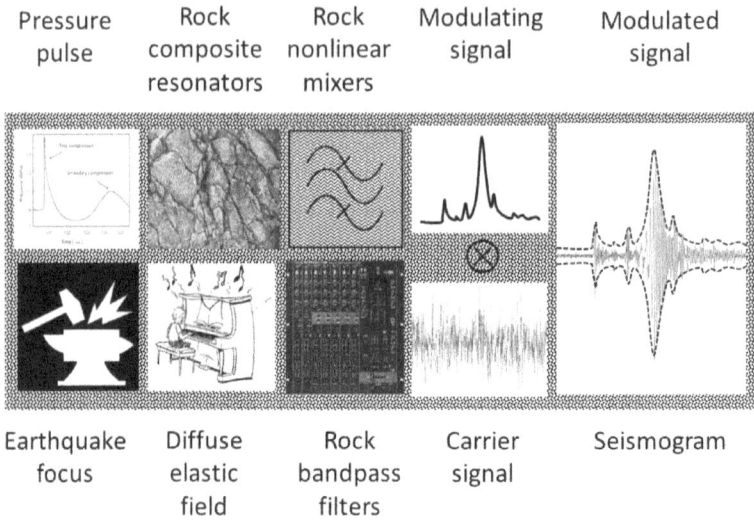

Pressure pulse · Rock composite resonators · Rock nonlinear mixers · Modulating signal · Modulated signal

Earthquake focus · Diffuse elastic field · Rock bandpass filters · Carrier signal · Seismogram

Fig. 1.1.1.9. Earthquake's equivalent circuit.

some are high-amplitude nonlinear waves that will form coda and low-frequency qP- and qS-waves that will travel long distances helped by the dispersion and nonlinearity of the medium of propagation. If the Earth were an isotropic, nondispersive, homogeneous continuum there would be no seismic waves; the pressure pulse will keep its original shape bouncing back and forth from various boundaries of the composite resonator systems until the whole energy gets dissipated. The dispersive and heterogeneous structure of the Earth composed by various rocks with diverse elastic properties form a large set of resonators able to generate from a single pressure energy pulse a whole spectrum of frequencies and waves—coda, created after passing the ground bandpass filters. Coda is composed by a great number of amplitude-, frequency-, and phase-modulated waves. Not all of these waves will take part of the earthquake or arrive to the seismic stations where the seismographs are located. Most of the waves will dissipate or scattered away in the rock. Many will continue their propagation longer time depending on various materials factors. Earthquakes usually last short time—a couple of seconds to one minute only. There could be aftershocks but the main shock is usually short. This is due to the 'rock bandpass filters' that let go only some waves and stop most of the rest. In this time window scattering processes will filter and shape modulating qP-, qS-, and surface waves that will be recorded on the seismograms together with modulated carrier coda waves.

An interesting effect is observed in Fig. 1.1.1.10 vertical velocity seismogram registered during the 2001 Tacoma-Olympia earthquake. The LR-wave time position is coming much earlier than the tail waves which

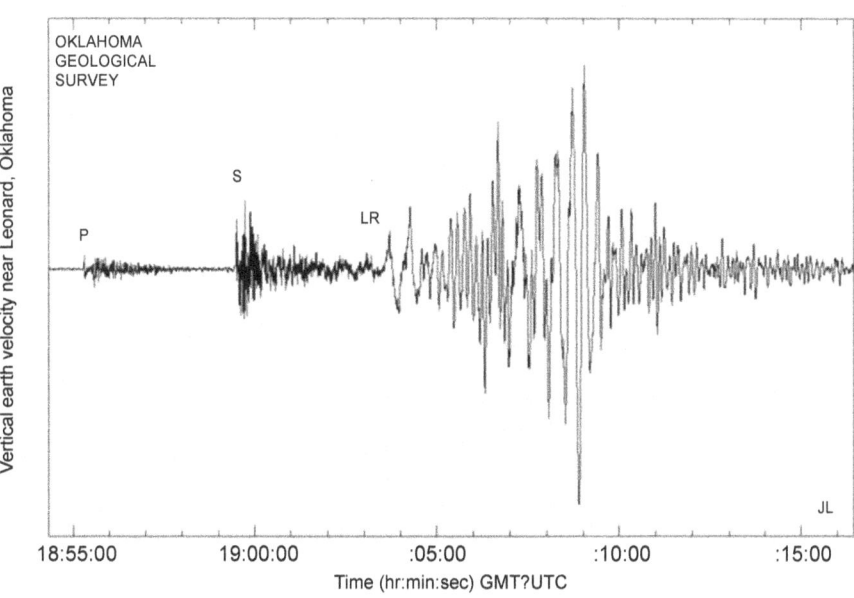

2001 Feb 28, Tacoma-Olympia earthquake, Ms = 6.9(OGS)

Fig. 1.1.1.10. 2001 February 28 Tacoma-Olympia earthquake seismogram.

vertical velocity is much higher. A possible explanation would be that this is a modulated surface LR-coda that was generated somewhere near to the location of the recorder due to self-modulation.

1.1.2 Self-modulation of seismic waves

In electronics and telecommunications modulations of the wave amplitude, frequency, or phase are used for the transportation of information. The electromagnetic wave 'carrying' the information is called carrier. An information-bearing electromagnetic wave is used to change some of the carrier wave parameters—frequency, amplitude, or phase, so the carrier will «carry» the information. At the other end the information is extracted from the modulated carrier by demodulation.

Amplitude modulation (AM) means that the amplitude of the carrier wave is changed by a modulating information-bearing wave that defines the envelope of the carrier waveform. In the frequency domain the amplitude modulation creates a wave which power is concentrated at the carrier frequency and two adjacent sidebands. Each sideband is equal in bandwidth to that of the modulating signal. In Fig. 1.1.2.1 are shown a synthetic carrier wave that is modulated by another modulating wave which has a longer period

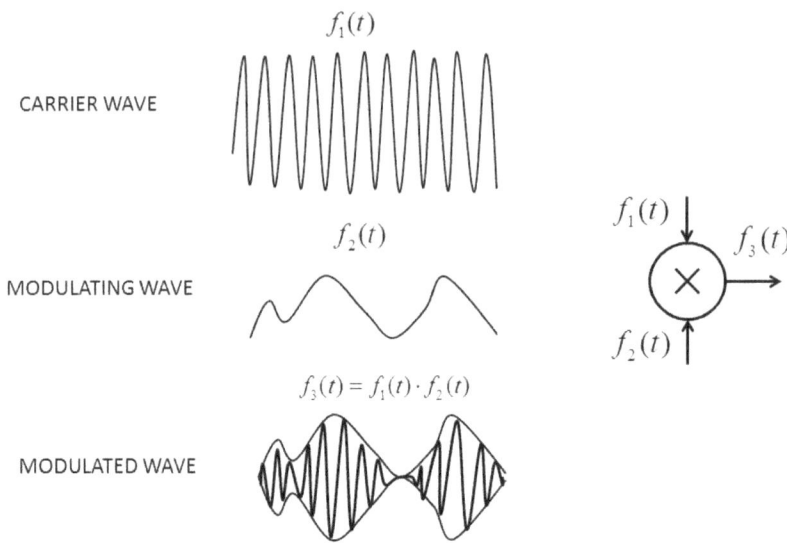

CARRIER WAVE $f_1(t)$

MODULATING WAVE $f_2(t)$

$f_1(t) \downarrow$ $f_3(t)$

$f_2(t) \uparrow$

$f_3(t) = f_1(t) \cdot f_2(t)$

MODULATED WAVE

Fig. 1.1.2.1. Amplitude-modulated elastic wave.

(or wavelength). In Fig. 1.1.2.2 is shown a seismogram and its bandpass and lowpass components. The carrier coda has been filtered and only the modulating waves are left.

It seems that the waves that have the greatest impact during an earthquake are the coda because they carry most of the seismic energy. The modulating P- and S-waves seems only to define when and in what form the coda energy will be delivered—P-wave first, followed by the S-wave, followed by surface elastic waves. However if the modulating waves overmodulate coda the amplitudes of both modulating and modulated waves increase as shown in Fig. 1.1.2.3. The coda of the surface seismic wave modulator looks different because it is a surface wave coda. A typical amplitude modulation seismograms registered during 2010 Haiti earthquake is shown in Fig. 1.1.1.8. In some areas overmodulation parts can be seen. P-, S-, LQ-, and LR-waves have been identified according to the arrival times, however, if various wave packets shown in the seismograms result from modulations of carrier waves such identification risks to be incorrect without studying the nature of the modulated coda carriers. This is especially valid in the cases of Love and Rayleigh waves because the velocities of propagation of these waves can vary in a wide range depending on the geological characteristics. Love waves are dispersive waves which velocity of propagation depends on frequency, thickness of the layered waveguide, and the ratio between the velocities of the SH waves in the layer and substrate. Rayleigh waves are nondispersive waves that do not require layered structures to propagate. However, layered

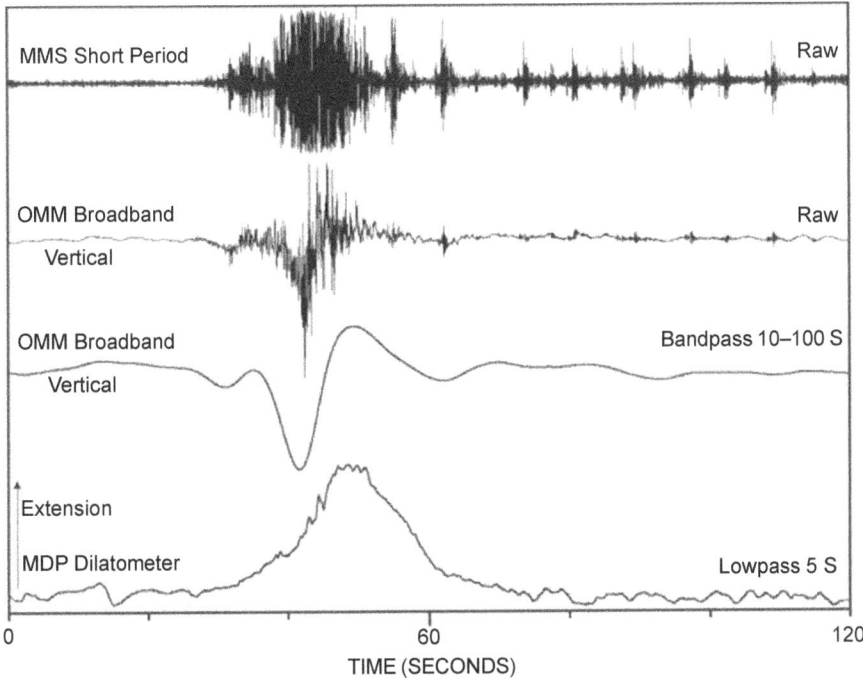

Fig. 1.1.2.2. Filtered seismogram (Mammoth Mountain).

structures they become dispersive and their velocity depends on the layer thickness and its elastic properties. Assuming that Love waves are always faster than Rayleigh waves could lead to errors without studying the specific conditions.

Some seismograms show recordings with well pronounced P-wave, S-wave, and L-waves as it is the case of the seismogram of the vertical displacement velocity (Fig. 1.1.1.10) recorded during the 2001 Tacoma-Olympia earthquake. In Fig. 1.1.2.2 are shown seismograms of the August 13, 2000, very-long-period earthquakes beneath Mammoth Mountain where it is hard to identify the type of the seismic waves. The second seismogram (Raw) shows a well defined wave packet built of a modulating wave pulse full of carrier coda. The next seismogram shows the band-passed pulse only where high-frequency coda has been cut out off by the filter. At the top is unfiltered trace from the short-period (1 second, vertical seismometer) Northern California Seismic Network on top of Mammoth Mountain (note the spasmodic burst). Middle two traces are unfiltered and filtered (10 to 100 second bandpass) vertical component records, respectively, from the CMG-3 broadband station (OMM) installed in a shaft of the Old Mammoth Mine (4 km (2.5 miles) southeast of Mammoth Mountain) and operated by the University of Nevada, Reno.

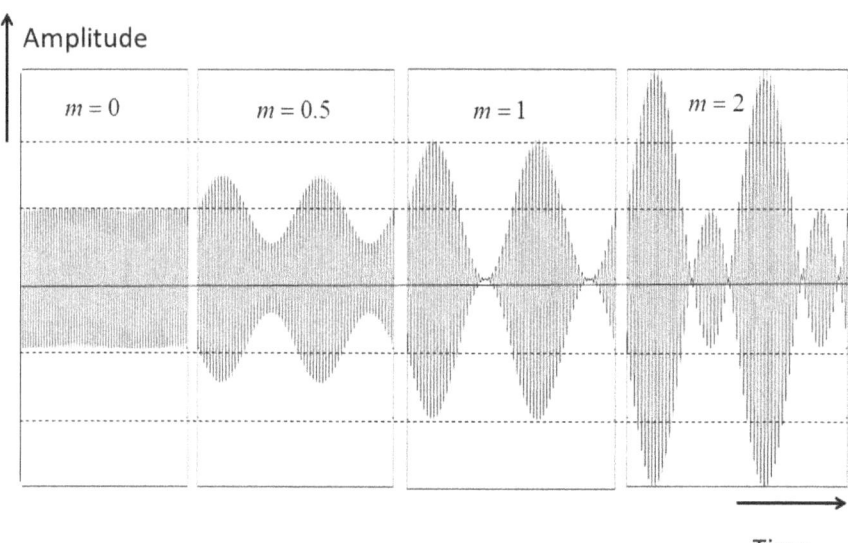

Amplitude

| $m = 0$ | $m = 0.5$ | $m = 1$ | $m = 2$ |

Time

Fig. 1.1.2.3. Modulation at 0% (left), 50% (second), 100% (third), and overmodulation 200% (right).

Bottom is a low-pass (5-second cutoff) trace from the POP borehole dilatometer located 4 km (2.5 miles) due west of the Mammoth Mountain summit. These examples demonstrate that the modulating signals are not always regular P-wave and S-wave. It looks like more complex filtering systems in the Earth are modulating coda. Coda waves could be subject to some amplification also due to local resonance effects or to a nonlinear amplification.

How does the amplitude modulation process work? Amplitude modulation consists in multiplication of waves (Frenzel 2008). Let us consider two sinusoidal waves, one called modulating and the other called modulated or carrier, with amplitudes A_c and A_m and angular frequencies ω_m and ω_c, respectively.

$$a_c(t) = A_c \sin \omega_c t$$
$$a_m(t) = A_m \sin \omega_m t \tag{1.1.2.1}$$

They may travel together and nothing impressive would happen. The two waves simply add to each other without any interaction (Fig. 1.1.2.4):

$$a_c(t) + a_m(t) = A_c \sin \omega_c t + A_m \sin \omega_m t \tag{1.1.2.2}$$

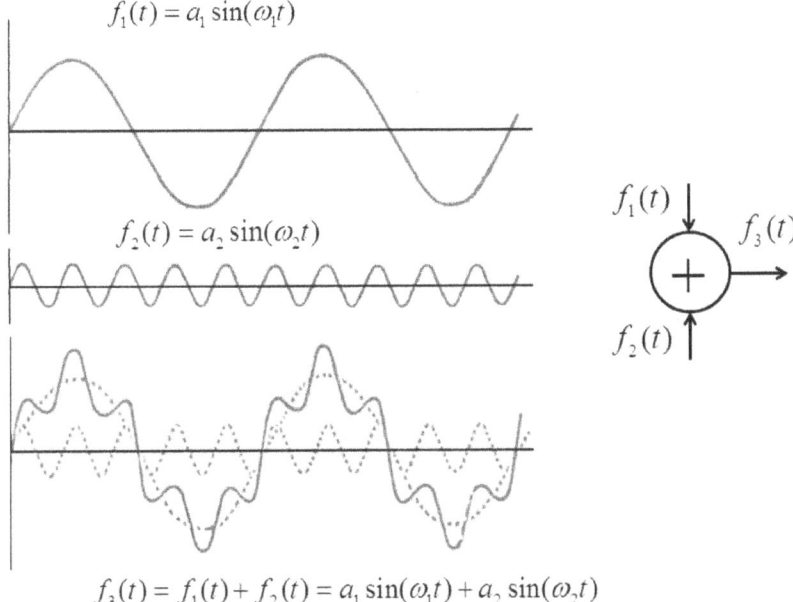

$$f_1(t) = a_1 \sin(\omega_1 t)$$

$$f_2(t) = a_2 \sin(\omega_2 t)$$

$$f_3(t) = f_1(t) + f_2(t) = a_1 \sin(\omega_1 t) + a_2 \sin(\omega_2 t)$$

Fig. 1.1.2.4. Summed waves.

However, trigonometry we know that:

$$\sin x \cdot \sin y = \frac{1}{2}\left[\cos(x-y) - \cos(x+y)\right] \qquad (1.1.2.3)$$

Using Eq. 1.1.2.3 we can rewrite Eq. 1.1.2.2 in the form:

$$a_c(t) \cdot a_m(t) = \frac{A_c A_m}{2} \cos\left[(\omega_c - \omega_m)t\right] - \frac{A_c A_m}{2} \cos\left[(\omega_c + \omega_m)t\right] =$$
$$\frac{A_c A_m}{2}\left\{\cos\left[(\omega_c - \omega_m)t\right] - \cos\left[(\omega_c + \omega_m)t\right]\right\} \qquad (1.1.2.4)$$

Now we have a new phenomenon which is very different from the simple addition of waves traveling together in Eq. 1.1.2.2. Here with waves with frequencies equal to the difference and the sum of the original waves we get multiplication of waves (Fig. 1.1.2.1). This is the basic process that has been used for many years in the analog radios for sending and receiving information using a device called mixer. If we want to transmit a 5-kHz audio tone by radio, we send it into one of the mixer's inputs and send an RF signal of, for example, 4555 kHz to the other mixer's input. At the output of the mixer we get two radio signals—one at 4550 kHz and another at 4560 kHz. These two signals can be easily demodulated using again a mixer. We send the signals to one of the input of the mixer, and another RF signal of 4555 kHz to the other mixer's input. At the output we will get our 5-KHz tone. The first mixer is the modulator and the second one—the demodulator. The analog signal AM modulation in our example is presented in Fig. 1.1.2.5. Above is an audio signal oscillogram looking very much like a seismogram of vertical displacement.

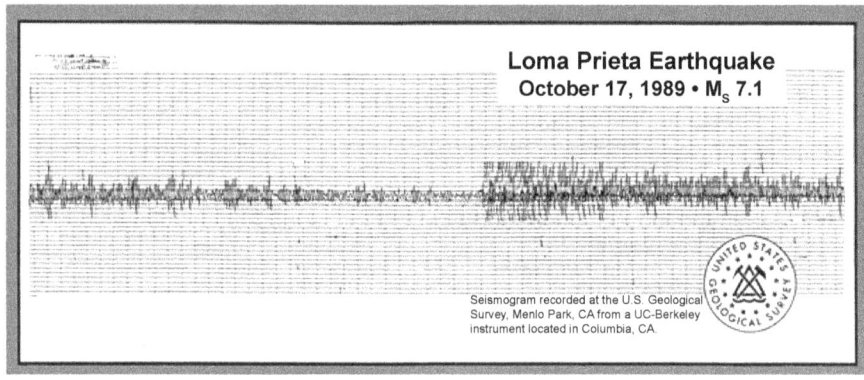

audio signal oscillogram

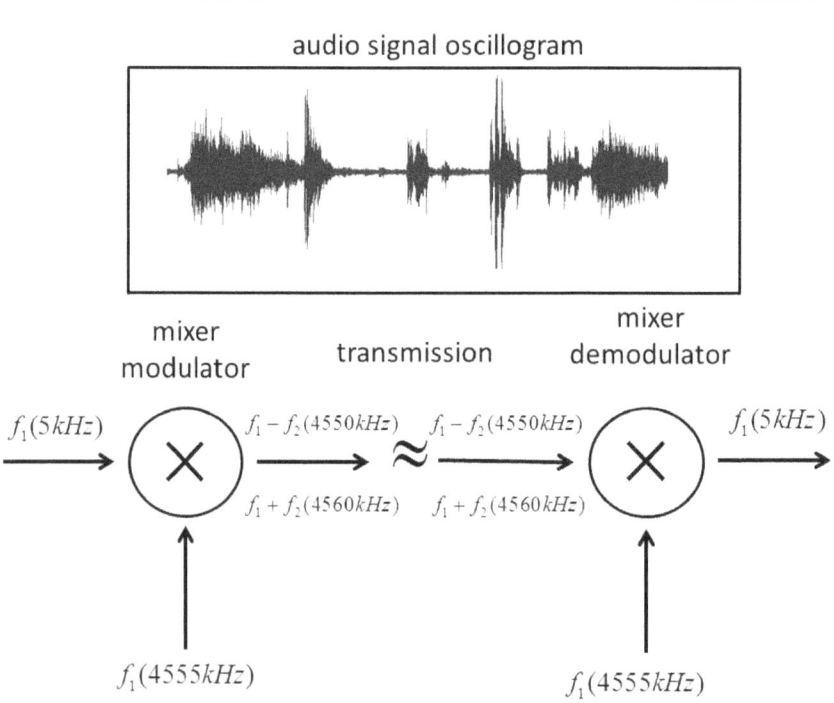

Fig. 1.1.2.5. Above—A seismogram from 1989 Loma Prieta earthquake looks as unmodulated coda. Middle—An audio signal oscillogram looks very alike seismograms of vertical displacement as ones shown in Fig. 1.1.1.7 and 1.1.1.8. Below—An analog signal of 5 kHz modulates a 4555 kHz carrier wave fed to a mixer from an oscillator and retrieved after demodulation by a second mixer.

Usually in radio communication the amplitude of the modulating signal (wave) is smaller than the amplitude of the carrier $A_m < A_c$ in order to be avoided signal distortions. The ratio:

$$m = \frac{A_m}{A_c}$$
(1.1.2.5)

is called modulating coefficient or degree of modulation. Figure 1.1.2.2 shows modulation at various values of the modulating coefficient. In the case of overmodulation the signal gets distorted and the amplitudes of both modulating and modulated signals increases. In the case of seismic waves the self-modulating process does not occur everywhere—only in areas with appropriate nonlinearity and dispersion where sesimic waves of combination frequencies grow followed by seismic beat. Figure 1.1.2.6 shows rail road damage by a localized self-modulation of a surface seismic wave. In the area where the modulated wave packet grew the road got a twist while the rest is slightly or not all damaged. It seems that the twist was done by a surface wave with a sidewinding displacement in the horizontal plane. Usually such displacement is associated of Love waves, however, we will see that other type

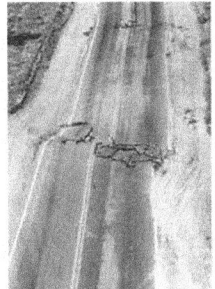

Fig. 1.1.2.6. Road damage caused by a localized self-modulation of a surface seismic wave with displacement in the horizontal plane—above left (Byers 1976), above right (Teasdale 2010), below left (National Geophysical Data Center), below right (Solar Roadways).

of surface elastic waves with similar to Love waves behavior can be generated in nonlinear conditions called skimming waves. Skimming waves can occur more often than Love waves because they do not require specific dispersion conditions as Love waves. Values for the carrier signal and the modulating signal can be used in a formula to express the complete modulated wave. First, we keep in mind that the peak value of the carrier is the reference point for the modulating signal; the value of the modulating signal is added to or subtracted from the peak value of the carrier. The instantaneous value of either the top or the bottom voltage envelope can be computed by using the equation:

$$a_1(t) = A_c + a_m(t) = A_c + A_m \sin(\omega_m t) \tag{1.1.2.6}$$

Therefore the instantaneous value of the complete modulated wave is equal to:

$$a_2(t) = a_1(t)\sin(\omega_c t) = \left[A_c + A_m \sin(\omega_m t)\right]\sin(\omega_c t) \tag{1.1.2.7}$$

Using m Eq. 1.1.2.7 can be presented in the form:

$$a_2(t) = A_c\left[(1+m)\sin(\omega_m t)\sin(\omega_c t)\right] \tag{1.1.2.8}$$

It seems that our examples are related to linear phenomena only. Indeed the mixers are designed to be linear devices. However, the amplitude modulation exists no matter whether the system is linear or not if the carrier and the modulator propagate in a nonlinear way. The mechanism underlying multiplication in any system (electric, elastic, or optic) is that nonlinearly distorts waveforms, causes combination frequencies $\omega_1 \pm \omega_2$, and acts as a multiplier. Linear wave propagation does not produce waves at combination frequencies $\omega_1 \pm \omega_2$. Seismic wave propagation is a strongly nonlinear process that produces new waves at combination frequencies which interactions with each other shape the characteristics of every earthquake.

The waveform of a non-sinusoidal signal can be changed by passing through a system that has only linear distortion, but only nonlinear distortion can change the waveform of a simple sine wave. It can also produce an output signal whose output waveform changes as a function of the input amplitude, something not possible with linear distortion. We will see this effect when we discuss the nonlinearity of rocks. Nonlinear systems often distort excessively with overly strong signals. Nonlinear distortion may take the form of harmonic distortion, in which integer multiples of input frequencies occur, or intermodulation distortion, in which different components multiply to make new ones. Multiplying a signal by itself generates harmonic distortion by adding the signal's frequency to itself. The amplitude modulation is a frequency-shifting system in which the original unmodulated signal traditionally called the carrier emerges from the mixer along with the sum and difference products, traditionally called sidebands.

A linear system (Fig. 1.1.2.4) is defined as:

$$F(a_1 + a_2) = F(a_1) + F(a_2) = F(A_1 \sin \omega_1 t + A_2 \sin \omega_2 t) \qquad (1.1.2.9)$$

Equation 1.1.2.9 shows that in a linear system no combination frequencies appear and all waves add to each other and travel unchanged together.

A nonlinear system (Fig. 1.1.2.1) is defined as a series of higher orders:

$$F(a_1 + a_2) = \beta_1 (a_1 + a_2) + \beta_2 (a_1 + a_2)^2 + \beta_3 (a_1 + a_2)^3 + ...$$

$$\beta_1 (a_1 + a_2) = \beta_1 \left[A_1 \sin(\omega_1 t) + A_2 \sin(\omega_2 t) \right] \qquad (1.1.2.10)$$

$$\beta_2 (a_1 + a_2)^2 = \frac{\beta_2}{2} \left\{ \begin{array}{l} 2A_1 A_2 \left[\cos(\omega_1 - \omega_2)t - \cos(\omega_1 + \omega_2)t \right] + \\ A_1^2 \left[1 - \cos(2\omega_1 t) \right] + A_2^2 \left[1 - \cos(2\omega_2 t) \right] \end{array} \right\}$$

....

The combination frequencies $\omega_1 \pm \omega_2$ appear in the nonlinear system describes by Eq. 1.1.2.10. We have mentioned that mixers are designed to be linear devices. However, if nonlinear signals go through them combination frequencies appear. Combination frequencies appear also if linear signals are fed through a nonlinear system such as transistor or diode in an electronics circuit. In the case of seismic waves we have both—nonlinear elastic waves propagating through nonlinear rock. Therefore, wave multiplication will always take place in seismic systems.

An important factor regarding seismic waves and amplitude modulation is overmodulation. The modulation limit is reached when the sum of the sidebands and carrier at the modulator output reaches zero at the modulating waveform's most negative peak. This situation is called 100% modulation Fig. 1.1.2.3. One-hundred-percent modulation (modulating coefficient 1) is a limit because an amplitude modulator cannot reduce its output to less than zero. Trying to increase modulation beyond the 100% point results in overmodulation in which the modulation envelope no longer mirrors the shape of the modulating wave shown in Fig. 1.1.2.1. If the unmodulated waves' amplitudes is calibrated to be equal to unit, 50% modulation increases the amplitude of both carrier and modulating envelop up to 1.5 times the unmodulated wave amplitude, 100% modulation brings the modulated amplitudes up to 2 times the unmodulated wave amplitude, and 200% overmodulation not only distort the waveforms, but also increases the amplitudes up to 3 times the unmodulated amplitude (Fig. 1.1.2.3).

Nonlinear effects of modulation in stability of seismic waves have been studied in numerical experiments (Pavlenko 2007). The results show that self-modulation and generation of harmonics during the propagation of high-amplitude seismicwaves occur if resonance intensification in soil layers take place and cause further increase of the wave amplitude. The phenomena of self-modulation and amplification of seismicwaves are caused by dispersion of propagation velocities in subsurface soils due to their nonlinear response.

1.1.3 Fourier analysis

Fourier analysis has been developed initially for analyzing periodic in time function (Fourier series). Later the Fourier analysis has been extended for analyzing nonperiodic function (Fourier transform). It is possible to go from Fourier series to Fourier transform by considering nonperiodic function as a limiting case of periodic function with a period tending to infinity. A discrete set of frequencies in the periodic case becomes a continuum of frequencies in the nonperiodic case—a spectrum, which make possible to analyze a signal in the time or space domain or in the frequency domain. A periodic function of time (for example a transient wave) can be presented as a sum of infinite number of simple sine waves—the Fourier series. The Fourier transform decomposes a function of time (not necessarily a periodic one) into the frequencies that make it up. The Fourier transform is the frequency domain representation of the original signal. The inverse Fourier transform (Fourier synthesis) combines all different frequencies to recover the original function of time. Most of the applications of the Fourier analysis were for solving the fundamental linear differential equations of physics (heat equation, wave equation, Laplace's Equation).

All waves—optic, elastic, electromagnetic, thermal, etc. are transient functions of time. The Fourier transform represents a transient in time wave as an infinite number of harmonics with frequencies $n\omega$ where $n = 1,2,3,...$ In a linear system it is sufficient to consider only one Fourier frequency component to find a general solution as a superposition of all Fourier components. In the case of finite-amplitude waves the linear approximation breaks down and nonlinear effects must be taken into account. Linear theory predicts exponential growth of unstable waves, but nonlinear effects cause saturation and limit the wave amplitude at a finite level.

Let's consider the pulse shown in Fig. 1.1.3.1. Since the energy pulse is very short in time it can be presented as a delta function of Dirac $\delta(t)$ equal to 1 in the time interval –0.5 to +0.5 sec and 0 outside of this interval. The Dirac delta function is:

$$\delta(t) = \begin{cases} +\infty, & t=0 \\ 0, & t \neq 0 \end{cases} \text{ and } \int_{-\infty}^{+\infty} \delta(t)\,dt = 1; \; \delta(0) = +\infty \qquad (1.1.3.1)$$

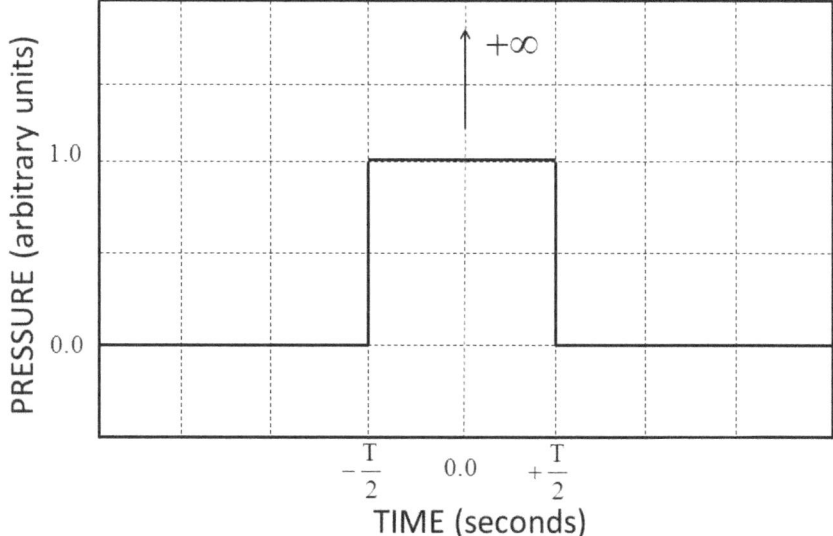

Fig. 1.1.3.1. Unit pressure pulse.

For any function $f(t)$ continuous at the point $t = t_0$ the Dirac delta function at the point $t = t_0$ is equal to the value of the function $f(t)$ at the point $t = t_0$:

$$f(t_0) = \int_{-\infty}^{+\infty} f(t)\delta(t - t_0)dt = \delta_{t_0}(f) \qquad (1.1.3.2)$$

According to Eqs. 1.1.3.1 and 1.1.3.2 can be presented also as:

$$f(t_0)\int_{-\infty}^{+\infty} \delta(t - t_0)dt = f(t_0) \qquad (1.1.3.3)$$

Following Eq. 1.1.3.2 the Dirac delta function can be presented as a superposition of an infinite number of harmonic functions with frequency v and amplitude 1. We remember that any twice differentiable function is a harmonic function. Therefore:

$$\delta(t) = \int_{-\infty}^{+\infty} e^{i2\pi vt}dv \qquad (1.1.3.4)$$

Harmonic functions form an orthonormal basis in the space of integrable functions and any physical signal can be decomposed on this basis, i.e., it can be presented as a sum of an infinite number of waves:

$$p(t) = \int_{-\infty}^{+\infty} P(t)e^{i2\pi vt}dt \qquad (1.1.3.5)$$

To determine $P(t)$ we can proceed with the integration of:

$$I(f) = \int_{-\infty}^{+\infty} p(t)e^{-i2\pi vt} dt \qquad (1.1.3.6)$$

From the Dirac delta function we get:

$$\int_{-\infty}^{+\infty} e^{i2\pi(v-f)t} dt = \delta(v-f) \qquad (1.1.3.7)$$

Therefore we have:

$$I(f) = \int_{-\infty}^{+\infty} P(f)\delta(v-f)dv = S(f) \qquad (1.1.3.8)$$

Using the variable v we get:

$$P(v) = \int_{-\infty}^{+\infty} p(t)e^{-i2\pi ft} dt \qquad (1.1.3.9)$$

The Fourier transform of the pulse in Fig. 1.1.3.1 is the *sinc* function $\dfrac{\sin \pi v}{\pi v}$ presented in Fig. 1.1.3.2.

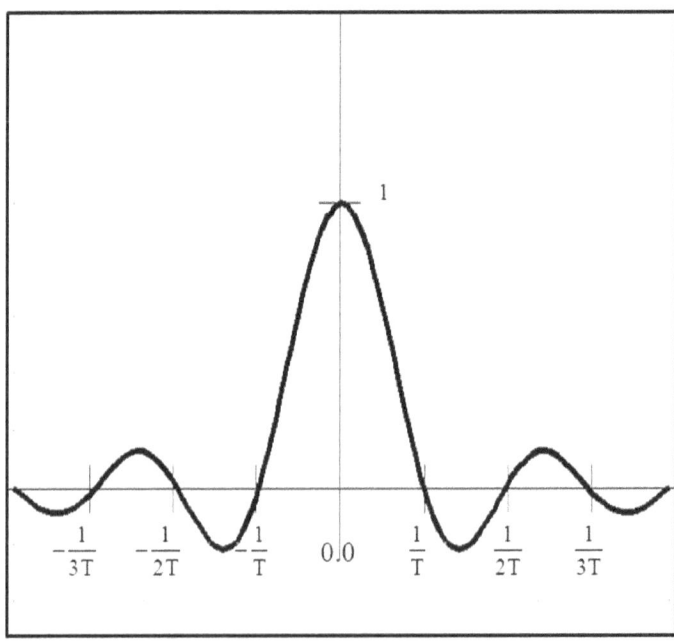

Fig. 1.1.3.2. Fourier transform of a pulse function.

If a series of pulses repeating each other at equal intervals of time are generated in the earthquake's focus they form a periodical function. According to Fourier's theorem any periodic or semi-periodic function can be presented as a sum of sines and cosines—a Fourier series. Each sine and cosine comes with their own amplitude and phase—Fourier coefficients.

$$f(t) = a_0 + \sum_{m=1}^{\infty} a_m \cos\left(\frac{2\pi m t}{T}\right) + \sum_{n=1}^{\infty} b_n \cos\left(\frac{2\pi n t}{T}\right) \qquad (1.1.3.10)$$

In one-term approximation we have:

$$f(t) = a_0 = \int_0^T f(t)dt \qquad (1.1.3.11)$$

Considering the second term we have:

$$f(t) = a_0 + b_1 \sin\left(\frac{2\pi t}{T}\right) \qquad (1.1.3.12)$$

where b_1 is given by:

$$b_1 = \frac{2}{T}\int_0^T f(t)\sin\left(\frac{2\pi t}{T}\right)dt = \frac{\pi}{2} \qquad (1.1.3.13)$$

The Fourier's coefficients can be presented in general form as:

$$a_0 = \frac{1}{T}\int_0^T f(t)dt$$

$$a_m = \frac{2}{T}\int_0^T f(t)\cos\left(\frac{2\pi m t}{T}\right)dt \qquad (1.1.3.14)$$

$$b_n = \frac{2}{T}\int_0^T f(t)\sin\left(\frac{2\pi n t}{T}\right)dt$$

For a series of pulses the Fourier's coefficients are:

$$a_0 = \frac{1}{2}$$

$$a_m = 0; m = 1, 2, 3, \ldots$$

$$b_n = \frac{2}{\pi n}; n = 1, 3, 5, \ldots \qquad (1.1.3.15)$$

$$b_n = 0; n = 2, 4, 6, \ldots$$

In reality the pressure pulses are coming at random intervals of time. Thus they form a function of time that never repeats—an aperiodic function. Each aperiodic function can be considered as period function with infinite period. To apply the Fourier's theory again on these aperiodic pulses we can generalize the Fourier series into an integral—Fourier transform. The Fourier transform presents a space-time effect such as an aperiodic function of time into a frequency domain. The Fourier transform is similar to the Fourier series with the difference that it is an infinite sum of infinitesimal sinusoids. The Fourier transform is given by:

$$F\left[f(t)\right] = F(i\omega) = \int_{-\infty}^{+\infty} f(t)e^{-i\omega t}\,dt \qquad (1.1.3.16)$$

The reverse Fourier transform is:

$$F^{-1}\left\{F\left(i\omega\right)\right\} = f(t) = \frac{1}{2\pi}\int_{-\infty}^{+\infty} F(i\omega)e^{+i\omega t}\,d\omega \qquad (1.1.3.17)$$

From the Fourier transform we notice an important consequence: Short signals (wave packets) in the time domain occupy a wide bandwidth. Wide signals (wave packets) in the time domain occupy a narrow bandwidth.

Using Fourier series and Fourier transform we can make an important conclusion: periodic or aperiodic pressure pulses (or wave packets) can always be presented as elastic waves that propagate in a frequency bandwidth. The shorter the pulse (or wave packet) in the time domain, the wider the wave's frequency bandwidth is. The wider the pulse (or wave packet) in the time domain, the narrow is the frequency bandwidth.

We have mentioned that the Gaussian function's representing the wave packet envelop Fourier transform is also a Gausian function. Indeed, if the normalized ($\int_{-\infty}^{+\infty} g(x)dx = 1$) Gaussian function is $g(x) = \dfrac{1}{\sigma\sqrt{2\pi}}e^{-\frac{x^2}{2\sigma^2}}$ where σ is the standard deviation (σ^2 is the variance) its Fourier transform is $G(\omega) = e^{-\frac{\omega^2\sigma^2}{2}}$, i.e., we get again a Gaussian function.

The spectral content of seismic waves varies significantly with time. They are non-stationary elastic waves. Fourier analysis is appropriate for stationary signals only. For non-stationary elastic waves we are interested in the frequencies that are dominant at any given time. Non-stationary elastic waves can be analyzed using short-time Fourier transforms. The short-time Fourier transform of a function at some time t is the Fourier transform of that function in a time window centered on t. For different time windows centered at different time the Fourier transforms are different. The short-time Fourier transform decomposes a signal into a set of frequency bands at any given time.

As mentioned already all twice continuously differentiable functions are harmonic functions. The harmonic functions form a complete orthogonal basis in the space of all integrable functions. That means any physical signal can be presented as a decomposition of an infinite number of harmonic functions (or waves):

$$s(t) = \int_{-\infty}^{+\infty} S(f) e^{i2\pi ft} df \qquad (1.1.3.18)$$

$S(f)$ represents the frequency spectrum of the signal.

Proceeding with the integration as we did in Eq. 1.1.3.18 we get:

$$S(f) = \int_{-\infty}^{+\infty} e^{-i2\pi ft} s(t) dt \qquad (1.1.3.19)$$

Equation 1.1.3.19 shows that the spectrum of a signal is its Fourier transform.

On the other hand, since any physical signal is a real function of time, we have:

$$S(-f) = \int_{-\infty}^{+\infty} e^{+i2\pi ft} s(t) dt = S^*(f) \qquad (1.1.3.20)$$

Equation 1.1.3.20 shows that the real part of the spectrum of the signal is pair and the imaginary part is odd:

$$\mathrm{Re}\left[S(-f)\right] = \mathrm{Re}\left[S(f)\right]$$
$$\mathrm{Im}\left[S(-f)\right] = -\mathrm{Im}\left[S(f)\right] \qquad (1.1.3.21)$$

Equation 1.1.3.21 takes us to the classic Euler's formula: $e^{ix} = \cos x + i \sin x$, where cos is a pair function, while sin is an odd function. For example, if we have a real physical signal with frequency f_0 $s(t) = a(t) \cos (2\pi f_0 t)$, which amplitude $a(t)$ is modulated in time, its spectrum $S(t)$ is its Fourier transform:

$$S(f) = \int_{-\infty}^{+\infty} A(t) \cos (2\pi vt) dt = \frac{1}{2} E(f - f_0) + \frac{1}{2} E(f + f_0) \quad (1.1.3.22)$$

The spectrum of $s(t) = a(t) \cos (2\pi f_0 t)$ is shown in Fig. 1.1.3.3.

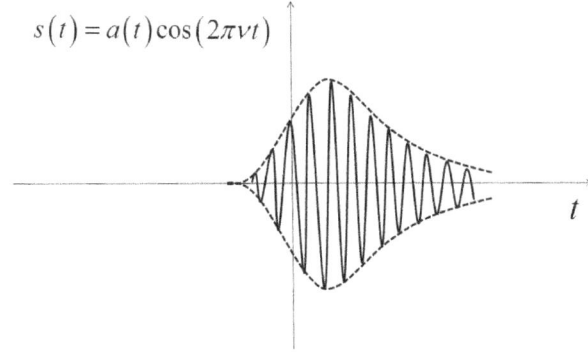

Fig. 1.1.3.3. The spectrum of $s(t) = a(t)\cos(2\pi f_0 t)$.

If instead a signal with an amplitude modulation we consider a signal with a frequency modulation by following the same procedure as above we get:

$$S(\omega) = \frac{1}{2}E(\omega - \omega_0) + \frac{1}{2}E^*(\omega + \omega_0) \qquad (1.1.3.23)$$

Equation 1.1.3.23 yields from $E(-\omega) = E(\omega)$. The spectrum of the pulse function shown in Fig. 1.1.3.1 is the *sinc* function presented in Fig. 1.1.3.2 and the Dirac δ-function's spectrum is 1.

We have demonstrated that the pressure pulse or a wave packet generate elastic waves. Now we will discuss the frequency response and pulse response of the medium of propagation.

A harmonic signal at the source can be presented as: $f_1(t) = e^{i\omega t}$. The response to this signal from the medium will be another signal delayed in time τ: $f_2(t + \tau) = e^{i\omega(t+\tau)}$.

According to Eq. 1.1.3.2 any function $f(t)$ can be presented in the form:

$$f(t) = \int_{-\infty}^{+\infty} f(t)\delta(t - \tau)\,dt \qquad (1.1.3.24)$$

Let assume that this function is generated at the source. We will be looking for the response of the rocks to that function. With the assumption that the medium is linear let suppose that the response to the Dirac's delta function is $\theta(t)$ and the response to $f(t)$ is the function:

$$g(t) = \int_{-\infty}^{+\infty} f(\tau)\theta(t - \tau)\,d\tau \qquad (1.1.3.25)$$

The function $g(t)$ is called convolution between the functions $f(\tau)$ and $\theta(\tau)$ or $f * \theta$. According to Eq. 1.1.3.24 the convolution between $f(t)$ and $\delta(t)$ is again $f(t)$. Therefore we have:

$$(f * \theta)(t) = \int_{0}^{t} f(\tau)\delta(t - \tau)\,d\tau = f(t) \qquad (1.1.3.26)$$

If $f(t)$ is a harmonic function $f(t) = e^{i\omega t}$ its response $g(t)$ according to Eq. 1.1.3.25 is equal to:

$$g(t) = \int_{-\infty}^{+\infty} e^{i\omega \tau}\theta(t - \tau)\,d\tau = e^{i\omega t}\int_{-\infty}^{+\infty} e^{-i\omega(t-\tau)}\theta(t - \tau)\,d(t - \tau) = f(t)H(\omega) \qquad (1.1.3.27)$$

The coefficient $H(\omega)$ in Eq. 1.1.3.27 is called transfer function:

$$H(\omega) = \int_{-\infty}^{+\infty} e^{-i\omega t}\theta(t)\,dt \qquad (1.1.3.28)$$

The Laplace transform is similar to the Fourier transform. The difference between them is that the Fourier transform represents a signal as a superposition of harmonic functions, whereas the Laplace transform represents a signal as a superposition of time moments. The Laplace transform is a transformation from the time-domain (all inputs and outputs are functions of time) to the frequency-domain (the same inputs and outputs are functions of frequency:

$$\Im(s) = \int_0^\infty f(t)e^{-st}\,dt \qquad (1.1.3.29)$$

The Laplace transform of $\delta(t-a)$ is:

$$\Im\big[\delta(t-a)\big] = \int_0^\infty e^{-st}\delta(t-a)\,dt = e^{-as} \qquad (1.1.3.30)$$

We note that the Laplace transform of the Dirac's delta function is 1:

$$\Im\big[\delta(t)\big] = 1 \qquad (1.1.3.31)$$

Therefore, the transfer function Eq. 1.1.3.28 is the Laplace transform Eq. 1.1.3.30 of the impulse response:

$$H(\omega) = \Im\theta(t) \qquad (1.1.3.32)$$

Any linear system is completely characterized by its impulse response. That means for any input the output can be calculated in terms of the input and the impulse response. The impulse response of a linear transformation is the image of Dirac's delta function under the transformation. The impulse response and frequency response are very useful for characterizing linear time-invariant (LTI) systems. The system is linear, so it obeys the principle of superposition. Stated simply, if two signals are linearly combined and input them to the system, the output is the same linear combination of what the outputs would have been had the signals been passed through individually. The system is time-invariant, so its characteristics do not change with time. A system's impulse response is defined as the output signal that results when an impulse is applied to the system input. It allows us to predict what the system's output will look like in the time domain. In a linear and time-invariant system if we decompose the input signal into a sum of components, then the output is equal to the sum of the system outputs for each of those components. If the input signal is decomposed into a sum of impulses the output would be equal to the sum the impulse responses scaled and time-shifted in the same way as the input impulses. An LTI system's frequency response provides a similar function: it allows you to calculate the effect that a system will have on an input signal, except those effects are illustrated in the frequency domain.

The impulse response of the square pulse in Fig. 1.1.3.1 as input to a LTI system is its Fourier transform presented in Fig. 1.1.3.2. Impulse response of the Bridgewater concert Hall to a single audio pulse is presented in Fig. 1.1.3.4. It is interesting to compare this data to the freeze-thaw resistance of concrete carried out by the Federal Highway Administration presented in Fig. 1.1.3.5. The pulse is input manually into the block. It is interesting to compare the seismogram in Fig. 1.1.1.5 to the impulse responses in Figs. 1.1.3.4 and 1.1.3.5.

If the rock were a LTI system the impulse or frequency response would not change with time. The frequency response of the rock would be the spectrum of its impulse response, which is also the reverse Fourier transform of the transfer function. Since the rock is a nonlinear medium this is not the case as shown.

An important application of the Fourier transform has been for long time the construction of solutions of partial differential equations describing processes evolving in time from a given initial state as for example an elastic wave which waveform gets distorted during its propagation starting from a sinusoidal wave. However, Fourier analysis applications are restricted mostly to linear differential applications. Generalized Fourier transform can be used for solving some nonlinear equations, but is cannot be extended to more complicated problems involving higher order derivatives. This presents a significant drawback for using the Fourier analysis tools for solving seismic wave propagation problems.

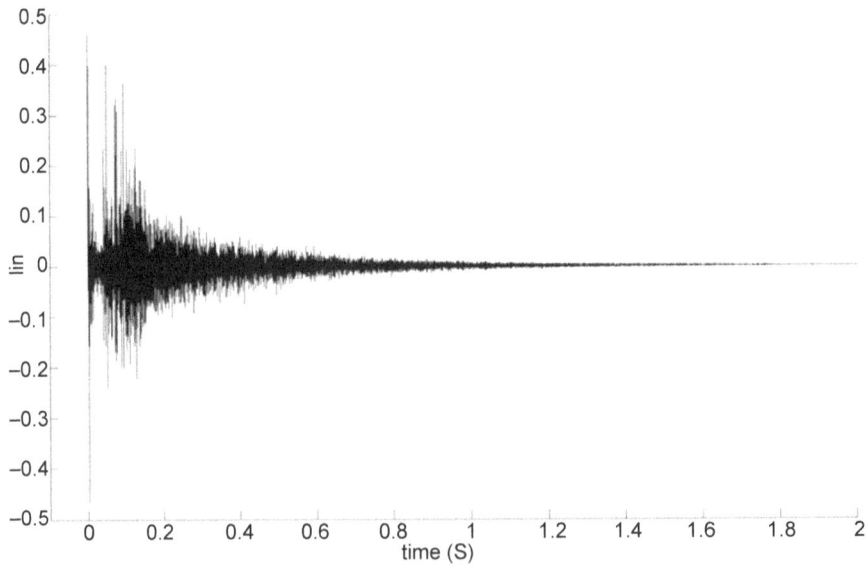

Fig. 1.1.3.4. Pulse response of the Bridgewater concert Hall.

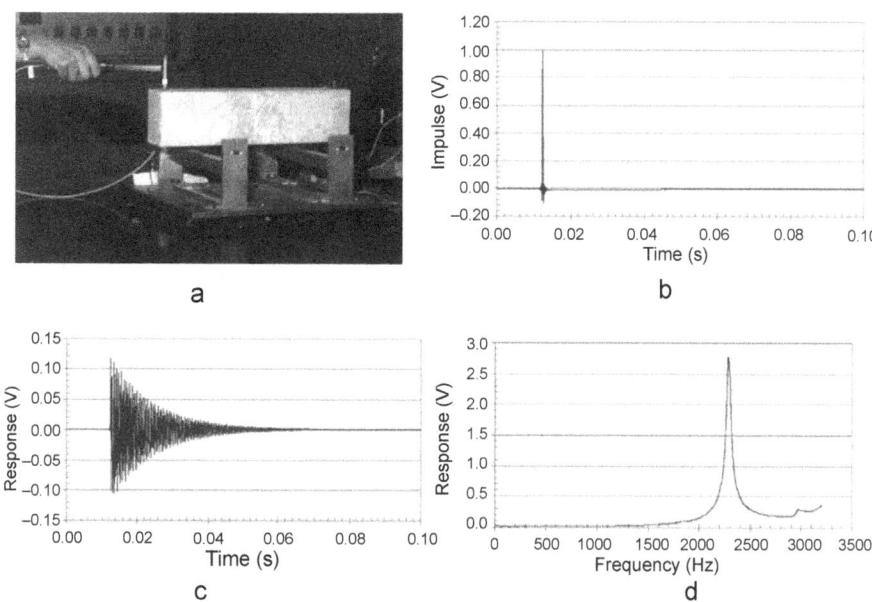

Fig. 1.1.3.5. (a) FHWA test setup; (b) Pulse (time domain); (c) Pulse response (time domain); (d) Pulse response (frequency domain) (FHWA 2006).

The Fourier transform is best to analyze stationary periodic functions which repeat themselves once per period without any modifications. The Fourier transform provides a single spectrum for the whole signal. For nonstationary waves frequency varies all the time, so at each moment it is important to know which frequency value is the dominant one. It is like to perceive a musical piece as a succession of single tones each one with its own spectrum, rather than as one big signal with an overall spectrum. In this case short-time Fourier transform is more appropriate.

Short-time Fourier transform of a function at time *t* is the Fourier transform of that function analyzed through some time-limited window centered at *t* (Basford et al. 1992). By sliding the examination window along in time will produce a set of Fourier that constitutes short-time Fourier transform. Short-time Fourier transform looks more appropriate for analyzing non-stationary elastic waves, but is has some weaknesses. If the examination window simply omits the signal outside the window, two problems are encountered. One is the sudden change in the power spectrum as a discontinuity enters or leaves the window, compounded by a lack of sensitivity to the position of the discontinuity within the window. The other problem is spectral leakage: if some component of the signal has a cycle time which is not an integral divisor of the window width, the transform exhibits spurious response at many frequencies. Also at high frequencies the number of waves in a window is high, producing good accuracy in frequency measurement; yet the window width prevents good localization of signal discontinuities, which the high frequencies otherwise could provide. Narrowing the window width to accommodate more precise time-localization of discontinuities causes other problems. A narrow window width is inappropriate at low frequencies, because a narrow windowing function spans fewer cycles. It distorts the signal noticeably over one wavelength, degrading accuracy of frequency measurement. Indeed, wavelengths longer than the window width cannot be measured. From these considerations it seems advantageous to let the windowing function be broad for analyzing low frequencies and narrow for high frequencies.

An example of windowing function is the Gaussian function $g(t) = e^{-at^2}$ The short-time Fourier transform at time τ can be expressed as:

$$F_g\left(t,\tau\right) = \int_{-\infty}^{+\infty} f\left(t\right)g\left(t-\tau\right)e^{-i\omega t}dt \qquad (1.1.3.33)$$

The response of the short-time Fourier transform centered at time $\tau = \tau_0$ to a pulse function $\delta(t - t_0)$ occurring at time $t = t_0$ is given by:

$$F_g\left(t,\tau_0\right) = \int_{-\infty}^{+\infty} \delta\left(t-t_0\right)g\left(t-\tau_0\right)e^{-i\omega t}dt = g\left(t-\tau_0\right)e^{-i\omega t_0} \qquad (1.1.3.34)$$

The power spectrum of the short-time Fourier transform is $F_g(\omega, \tau_0) = g^2$ $(t_0 - \tau_0)$ or the power spectrum is the same for all frequencies. The cross-section of the transform at constant frequency produces a time-reversed copy of the windowing function. Thus, the width (standard deviation) of the windowing function limits the accuracy with which the impulse can be located in time.

The equation for $F_g(t, \tau)$ shows that the short-time Fourier transform is an integral of the product of the function $f(t)$ with a set of basis functions $g(t - \tau)e^{-i\omega t}$ which vary over frequency and time τ. All basis functions have the same time-amplitude envelope (Fig. 1.1.3.6). The short-time Fourier transform decomposes a signal into a set of frequency bands at any given time.

A wavelet transform is similar to short-time Fourier transform. It also decomposes a signal into a set of frequency bands (called scales), by projection the signal onto a set of basis functions called wavelets. Projecting the signal onto different scales is equivalent to bandpass filtering. The wavelet basis functions are similar to each other. They vary only by dilatation and translation (Fig. 1.1.3.6). The wavelet transform basis functions are scaled in time to maintain the same number of oscillations and scaled in amplitude to maintain energy.

Short-time Fourier transform

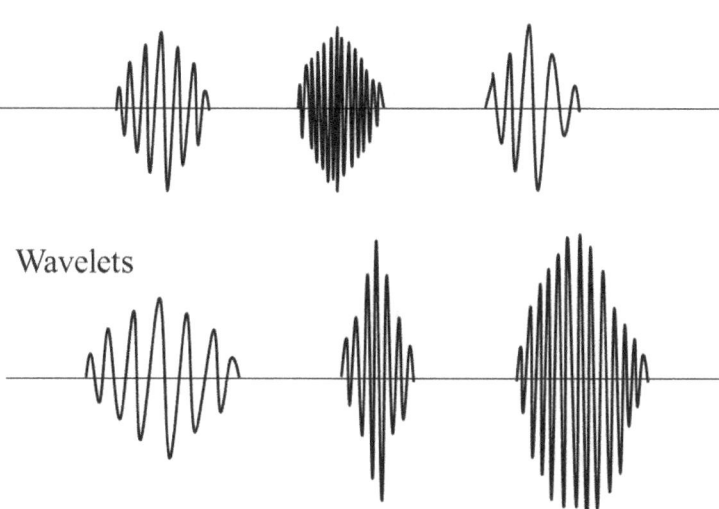

Wavelets

Fig. 1.1.3.6. Wavelets.

1.2 Seismic wave parameters and characteristics

The elastic wave polarization is defined by the direction along which the wave displaces the ground particles during its propagation (Fig. 1.2.1).

A longitudinal elastic wave displaces the ground particles along the direction of its propagation, while a shear elastic wave moves the ground particles along some of the directions that are perpendicular to the direction of propagation of the wave. Two independent shear waves and one longitudinal wave can propagate in each direction in an isotropic infinite solid. In plane-wave approximation (the medium of propagation and the wave-front radius are considered to be infinite) the two independent shear waves have flat fronts and orthogonal polarizations. Their polarizations form a plane that is perpendicular to the direction of propagation (and polarization) of the longitudinal wave. Both shear waves propagate at the same velocity (degenerated waves). In anisotropic solids the two shear waves have different velocities of propagation—one fast and one slow shear waves. Any elastic wave carries some energy characterized by energy density (the amount of energy per unit volume) and energy vector (vector of Poynting). In isotropic solids the energy vectors of all three waves are parallel to the direction of propagation.

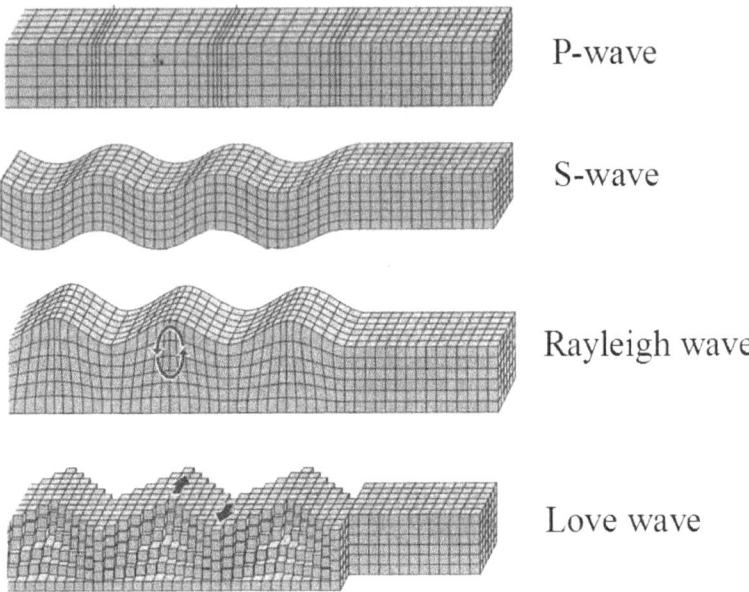

P-wave

S-wave

Rayleigh wave

Love wave

Fig. 1.2.1. Body P-wave and S-wave, Rayleigh and Love surface elastic waves.

In anisotropic solids this happens only in certain directions called pure-mode axes. In any other direction of propagation the wave vector and the vector of energy point in different directions. The polarizations are not orthogonal anymore—they have overlapping components and their projections on each other are not zero. These waves are not purely longitudinal or shear anymore—they are quasi-longitudinal and quasi-shear waves—qP-wave and qS-wave.

Many P- and S-waves reach the Earth's surface and get reflected back into the Earth's interior splitting their energy among new body waves with different velocities and polarizations. However, some of the elastic waves reach the ground surface under specific angles of incidence called critical angles and get converted into surface elastic waves. The surface waves continue their propagation keeping their energy closely confined to the ground surface. Usually surface elastic waves are the most dangerous ones to man-made contractions located on the Earth's surface because of their polarization and higher energy density. The specific conditions in which various types of surface elastic waves get created will be discussed later in this book.

The depth of the earthquake focus is an important parameter. In most earthquakes the hypocenter is located in the Earth's crust at 10–20 km. Such shallow-focus earthquakes causes usually more damage to man-made constructions on the Earth's surface than earthquakes with deep hypocenters located at hundreds of kilometers in the Earth's mantle. The epicenter of an earthquake is the point at the Earth's surface just above the focus of the earthquake. Deep earthquakes are not dangerous to man-made structures on the Earth's surface because they do not create surface seismic waves and do not cause aftershocks. The aftershocks are observed often during earthquakes with sallow focus because they either come directly from the hypocenter or they are caused by refracted waves.

If the earthquake's focus happens to be located under the ocean's bottom some P- and S-waves can reach the ground-water boundary. Depending on the angle of incidence these seismic waves could continue their propagation on the water-seabed interface as surface waves. In fluids (air, water) only P-waves can propagate, S-waves are forbidden. Refractions at the water-seabed boundary result in propagation in the water of P-waves only. The P- and S-waves reaching the water-seabed interface cannot cause a tsunami—a huge ocean wave that propagates on the surface across the ocean. Tsunami can be generated from an abrupt deformation of the seabed resulting in vertical displacement of huge volume of the overlying water. This abrupt deformation of the seabed will cause also an earthquake, so both tsunami and seismic waves are created independently. A tsunami can cause severe damages to the coastline as it happened to Fukushima, Japan, on March 11, 2011. On April 1, 1946, a magnitude of 7.8 on Gutenberg-Richter scale earthquake occurred near the Aleutian Islands, Alaska. A tsunami was generated also which inundated Hilo on the island of Hawai'i 3,000 km away from Alaska with a 14-metre high surge. Both tsunami waves were caused by the moving by 8–10 cm/year Pacific tectonic plate which was subducted and pushed downwards under the overriding North-American tectonic plate at the Ring of Fire under Alaska and under Honshu's undelaying plate in Japan.

When P- and S-waves cross boundaries between rock masses with different elastic properties the get refracted (transmitted) or reflected (sent back to the first medium of propagation). Depending on the wave's polarization and angle of incidence the reflected wave could have the same polarization and velocity of propagation as the incident wave, or it could change its polarization and velocity of propagation. The refracted wave which starts crossing the new medium of propagation can keep its polarization or change it, but its velocity of propagation will be different because the second medium has different elastic properties. During all these transformations of the wave the energy is conserved—the newly created elastic waves with different polarizations and velocities of propagation share the energy of the elastic wave they originate from. If the elastic energy dissipation during the propagation is low the elastic waves can travel long distances crossing various rock masses, converting their polarizations, changing their velocities of propagation, and splitting their energy. All these waves create an elastic field. This propagating strain-stress field is formed by bulk elastic longitudinal and bulk shears waves some of which could be registered by the seismic stations as aftershock P-waves and S-waves.

In anisotropic solids three types of elastic waves with different polarizations can propagate in each direction—one longitudinal (P-wave) with particles moving in the direction of propagation and two transversal or shear waves (S_1- and S_2-waves) with particles moving in directions perpendicular to the direction of propagation and also perpendicular to each other as shown in Fig. 1.2.1 for a flat-front wave (or plane-wave). The two S-waves propagate with different velocities and different wavelengths in an anisotropic medium and with the same velocity and wavelength in isotropic medium. The P-wave is always the fastest one followed by S_1 (fast shear) and S_2 (slow shear) in anisotropic medium. In an isotropic medium the two shear waves have the same velocity of propagation but the P-wave again is faster than them. In most cases the attenuation coefficient of the P-wave is slightly higher than the attenuation coefficients of the S-waves. In fluids such as air or water only P-waves can propagate. Shear waves are forbidden. Boundary conditions of a free space do not allow propagation of S-waves. As we will see later in this book that the absence of shear waves in fluids has an important consequence for earthquakes with epicenters in the ocean.

In isotropic medium the elastic waves could be assumed of being pure longitudinal or shear waves with their group-velocity vector (direction of energy transported by the elastic wave) and phase-velocity vectors (direction of propagation of the elastic wave) are parallel to each other. In anisotropic materials such as low-symmetry single crystals the phase and group velocities are parallel only in specific directions called pure-mode axes. In all other directions they are not parallel. The polarizations are not linear or orthogonal and the P- and S-waves are called quasi-longitudinal (qP-wave) and quasi-shear (qS-wave).

Various types of surface elastic waves exist. The most relevant to seismology are Love, Rayleigh, and Stoneley waves. The polarization of Love waves is always sidewinding—the particles are displaced in the plane of propagation of the surface wave perpendicularly to its direction of propagation. Rayleigh surface waves are rolling similarly to ocean waves. Their polarization is perpendicular to the direction of propagation and perpendicular to the surface of propagation. The vertical plane perpendicular of the surface of propagation is called sagittal plane. The ground particles follow elliptical orbitals in the sagittal plane.

The surface elastic waves' energy is confined close to the Earth's surface. Their amplitude decreases exponentially in the bulk under the surface. Sometimes large rock slabs located close to the Earth's surface can form boundary conditions that are similar to the boundary conditions in an acoustic waveguide. In these conditions Love waves are generated. The sidewinding

Love waves carry high energy density (W/m³) enclosed in the 'acoustic waveguide' formed by the rock slab because of the total internal reflection that occurs on the slab's boundaries. Only very small amounts of elastic energy could 'leak' to the outside 'cladding' of the rock 'core' so the wave can travel long distances in the waveguide keeping its energy density almost unchanged. Such guided surface waves are of the greatest danger to cities that happen to be located at their reach. Love waves require specific conditions to be created, namely a layered structure acting as a elastic waveguide. As we will see in Section 3.7.3, another type of Love-like surface waves exist in nonlinear dispersive media of propagation that do not require waveguide structures to grow and can be as destructive as Love waves. These waves are called skimming waves and they can grow only in nonlinear dispersive conditions which is probably the reason not to have attracted the attention of the seismological community. Figure 1.1.2.6 shows photos of railroad tracks near Bealville, California, where rails were bent and twisted by the 1952 Bakersfield earthquake caused by the White Wolf Fault (Photo: National Geophysical Data Center). Many other photos of twisted railroads have been posted online from other more recent earthquakes around the world. At instance, a couple of photos were posted by M. Teasdale after the September 4, 2010, earthquake in Cranbury, New Zeland (Teasdale 2010). Assuming linear propagation of surface seismic waves it is difficult to explain why only short segments of the railroad were twisted and not the entire railroad. It seems that the sidewinding force was caused by a horizontally polarized seismic wave. It is possible that a Love wave twisted the rails if some geological layered structure existed in that location forming an elastic waveguide. It is also possible that a skimming surface wave grew locally because of strong local nonlinearity and dispersion causing a seismic beat and localized self-modulation.

Rayleigh waves are generated in a different way. They do not need waveguide conditions to propagate. They are generated more often than Love waves and cause less destruction than Love waves. Rayleigh waves are complex surface waves composed of one longitudinal component and one shear component. These two waves are coupled. This means that they propagate always together and cannot be exist separately and propagate independently. The shear wave moves the ground particles along elliptical trajectories laying in sagittal plane and following the longitudinal wave. They are called often rolling waves and look very much as ocean waves. The ocean waves get their energy from the wind, while Rayleigh waves get their energy from bulk elastic waves coming to the ground surface under a critical angle.

The probability a bulk seismic wave to strike directly a city is much lower than the probability a surface elastic waves to do that. Bulk longitudinal and shear waves could strike and damage occasionally man-made constructions in the ground such as pipelines, tunnels, or water dams and cause various levels of damage depending on their polarization and power.

The frequency bands in physical acoustics are usually in the KHz, MHz, and GHz ranges. The seismic waves have much lower frequencies in the range from 0.01 Hz to 50 Hz. Human ear can detect acoustic waves (sound) from 20 Hz to 20 KHz. Dog's hearing frequency range is about three broader than human's one—up to 60 KHz. Cats can hear sounds up to 79 KHz while bats up to 100 KHz. Stories have been reported that animals can register the first seismic shakes of an upcoming earthquake and get agitated warning in this way humans. From the relation $\lambda = V/v = 2\pi V/\omega$ where λ is the wavelength, v is the frequency, $\omega = 2\pi v$ is the angular frequency, and V is the velocity of propagation of the seismic wave we can get some of the parameters. The acoustic wave velocity of propagation in solids is in the range between 2,000 m/sec and 8,000 m/sec. Therefore the wavelength is 5.10^{-5} m for $V = 5.10^3$ m/sec at frequency of 100 MHz. The acoustic wave velocity in the air is 330 m/sec, in water $-1,497$ m/sec at temperature of 25°C, and in granite -5.10^3 m/sec. These velocity values are for longitudinal waves only. Longitudinal waves move the particles of the medium of propagation along the direction of propagation. Only longitudinal waves can propagate in fluids (air, water). Shear waves (or transverse waves) are prohibited in fluids. Shear waves can exist only in solids. They move the particles of the medium of propagation perpendicularly to the direction of propagation. The direction of the displacement of the particles by an elastic wave is called polarization of the wave.

The seismic shock can last from hundreds of milliseconds to a couple of seconds. This means a row of elastic pulses are upcoming one after the other during this time period. In rare cases they can last longer—up to 30–60 seconds—often followed by numerous aftershocks. The seismic waves can travel long distances before they lose their power because of the low attenuation coefficients of elastic waves in the Earth's crust and mantle and also because of their low frequency. Longitudinal and shear bulk waves are always faster than surface elastic waves. For example, a bulk elastic wave with a velocity of propagation $V = 4,000$ m/sec and a frequency v in the range between 0.05 Hz and 50 Hz the wavelength λ is in the range between 80,000 m and 80 m. For a surface seismic wave propagating with a slower velocity of say 2,000 m the wavelength could vary between 40,000 m and 40 m. In physical acoustics usually frequency values are much higher corresponding to much shorter wavelengths than seismic waves. For instance, at 10 MHz a bulk waves with velocity of 4,000 m/sec will have a wavelength of 4.10^{-4} m (40 microns). The wavelength of a surface elastic wave propagating at 2,000 m/sec will be 2.10^{-4} m/sec. The ratio between the wavelengths of 1 Hz wave and 10 MHz is 10^7. This means that a 'microearthquake' can be simulated in laboratory conditions by generating acoustic waves of 10 MHz in a 0.01 m-side cube cut from a rock. This would correspond to 1 Hz seismic waves generated in a 10^5 m (100 km)-side cube of Earth's bulk.

1.3 The Earth's body structure

The continental Earth crust's thickness is about 40 km with a density estimated to be about 2.7×10^3 kg/m^3. The oceanic crust is thinner—5–10 km than the continental crust but it is denser—about 2.9×10^3 kg/m^3. The Earth's mantle is denser than the crust—3.3×10^3 kg/m^3. The temperature of the mantle close to the bottom of the crust is between 500°C and 900°C, while on the bottom close to the core it is about 4,000°C with a pressure of 136×10^9Pa. The velocity of propagation of a P-wave in the crust is about 6×10^3 m/sec while in the mantle it reached 13×10^3 m/sec. Most of seismic activities happen in the crust, however powerful bulk elastic waves could propagate in the Earth's mantle as well. The boundary conditions between the mantle and crust are such that the transmission of bulk waves from the mantle into the crust is easier than the other way about—from the crust into the mantle. This is due to the higher elastic impedance of the mantle $Z = \rho.V$ (ρ is the density than of the crust and V- velocity of propagation). As it will be discussed later in this book this means more elastic waves traveling through the crust will get reflected back to it at the boundary with the mantle. The most seismically active part is the lithosphere where most of the seismic waves propagate. The Earth's structure is shown in Fig. 1.3.1. The crust's chemical composition is mostly silica (Si$_2$O) 60%, alumina (Al$_2$O$_3$) 15%, iron oxide (FeO and Fe$_2$O$_3$) 7%, lime (CaO) 5%, magnesia (MgO) 3%, sodium oxide (Na$_2$O) 3%, potassium oxide (K$_2$O) 3% and the rest is titanium dioxide, phosphorous pentoxide, water, and carbon

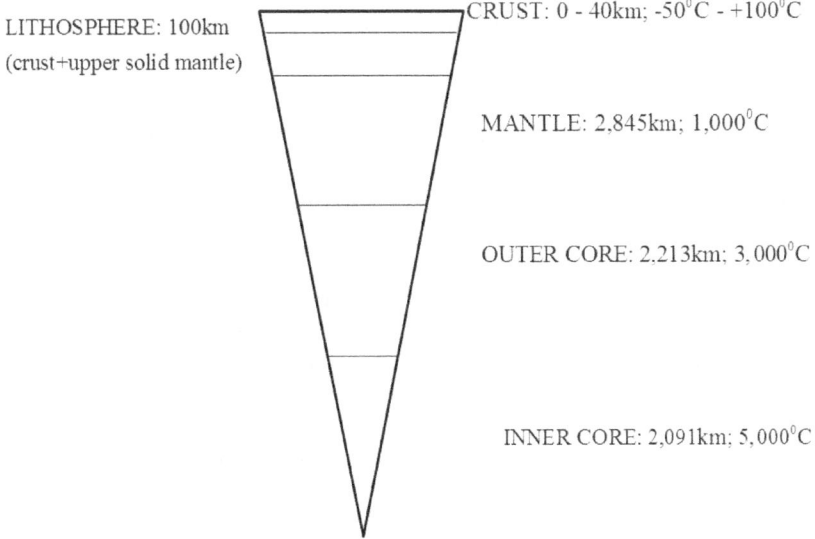

LITHOSPHERE: 100km
(crust+upper solid mantle)

CRUST: 0 - 40km; -50^{0}C - +100^{0}C

MANTLE: 2,845km; 1,000^{0}C

OUTER CORE: 2.213km; 3,000^{0}C

INNER CORE: 2,091km; 5,000^{0}C

Fig. 1.3.1. The Earth's structure.

dioxide. All these oxides have very low elastic wave attenuation coefficients in the frequency range 0.5–50 Hz.

The power loss per unit length of distance of propagation of elastic waves in solid media depends on many factors such as frequency, temperature, pressure, and attenuation coefficient of the medium of propagation. The heterogeneity of the Earth cause intense scattering of the elastic wave. Usually elastic waves can travel many kilometers before losing their power.

It has been established that most rocks exhibit a strong nonlinear elastic response. Non-linear harmonics generation has been also observed in laboratory experiments (Johnson and Shankland 1989). Experimental investigations show that the earth's crust is elastically nonlinear and contains the sources of accumulated elastic energy, so that it can be considered as a nonlinear active medium. The effects of stimulation of narrow- and broad-band seismic emission and the existence of dominant frequencies for which the medium is transparent have been reported. The temporal evolution of chaotic signals results in the formation of quasi-sinusoidal steady-state wave trains (Beresnev and Nikolaevskiy 1993).

Dispersion is an important phenomenon which has a significant impact on the seismic wave behavior and propagation. A pulse wave of a finite duration changes its waveform during propagation in a dispersive medium because its individual spectral components propagate with different velocities. Also resonance effects between high-order harmonics that are observed in nondispersive media do not occur in dispersive ones. Two types of dispersion can exist—dispersion associated with attenuation and scattering of the waves due to the structure of the medium of propagation, and dispersion which is due to the interference of the waves when they propagate in layered structures, waveguides, and free surfaces. Both types of dispersion can exist in the rocks.

In an isotropic elastic solid the energy transport velocity is equal to and in the same direction as the velocity of propagation of the elastic wave. In isotropic solids all directions are equal with respect to physical properties. All physical parameters involving some movement can be described with simple vectors. In anisotropic solids physical parameters vary depending on the direction in the space. Simple vectors cannot describe all physical parameters. In this case tensors are used to describe the physical properties. A vector is a tensor of first rank. In an anisotropic solid tensors of higher ranks are used to describe various physical properties. Heterogeneous solids are not anisotropic solids. They are composed of various domains with different physical properties. Some of the domains can be isotropic or anisotropic. If an elastic wave is propagating initially in an isotropic domain where all physical properties are described by simple vectors or scalars only crosses the boundary with an anisotropic domain where birefringence occurs and physical properties are described by tensors the elastic wave will be modified depending on its polarization. These modifications are very important especially for shear waves. It is interesting to discuss how anisotropy can appear in the Earth's crust. Actually anisotropy can be created in the crust through various mechanisms. For example an isotropic domain in the crust exposed to high pressure can become elastically anisotropic because its density in the direction of the pressure will be higher than in directions perpendicular to that direction. This will affect the velocity of propagation. There will be two different velocities in the direction of the pressure or perpendicularly to it. The velocity of the energy transport and the velocity of propagation of the elastic wave will differ by direction and by magnitude. Such induced seismic anisotropy is very similar to the elastic anisotropy of a ceramic cubic sample clamped in a vise.

The Earth's crust and mantle are neither purely isotropic nor anisotropic—they are heterogeneous. Both isotropic and anisotropic materials are homogeneous. They are uniform in composition and their physical properties do not change along any direction in the space. Physical properties of isotropic materials are not only the same along any direction in the space but they are also the same for all directions in the space. Physical properties in isotropic media can be described by vectors and scalars. Physical properties of anisotropic materials are the same along any direction of the space but they vary with the direction in the space. Physical properties in anisotropic media can be described by tensors of various ranks. A tensor of first rank is a vector. Figure 1.3.2 shows the optical indicatrices of fused quartz and trigonal

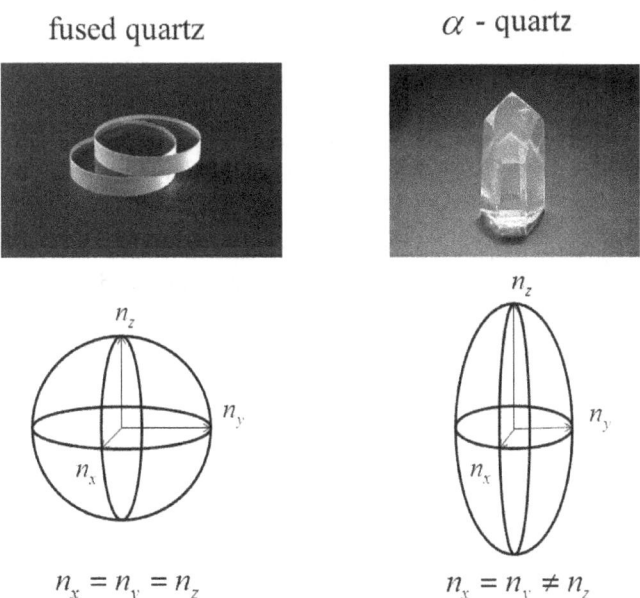

Fig. 1.3.2. Optical indicatrices of isotropic fused quartz and anisotropic trigonal crystal of α quartz.

single crystal of α-quartz. The optical indicatrix of fused quartz is a sphere with equal indices of refraction in all directions. The optical indicatrix of the α-quartz is ellipsoid with equal indices of refraction along the x- and y-axes and a different index along z-axis which is called optical axis.

The Earth's crust and mantle are neither purely isotropic nor anisotropic—they are heterogeneous. A heterogeneous material is nonuniform. It is composed by diverse domains that can be connected or separated by other domains all having different physical properties. Various rocks are heterogeneous in a different way. Rock salt is composed by anisotropic single crystal which are oriented randomly in the space which makes them as a whole to behave as an isotropic solid. Brecca is a typical heterogeneous material composed by fragments of minerals cemented by material with different elastic properties. Flint is similar to brecca but it is composed by gig chunks of mineral quartz with superior elasticity behaving as an isotropic solid surrounded by soft chalk or limestone which has very different elastic properties that mineral quartz. Pumice is a porous volcanic rock composed of hard volcanic glass which as a uniform solid has also superior elasticity.

All elastic waves that cross these domains with different elastic properties will be scattered or if transmitted they change their velocity of propagation. From $\lambda = V/v = 2\pi V/\omega$ that means that the wavelength λ changes depending on V at constant frequency $v = const$. However, in some cases the frequency is not constant; it changes with the velocity of the wave. When frequency changes when velocity changes it means that there is dispersion and the media is dispersive.

Here there is something important to notice. All these different domains that form a heterogeneous medium are randomly oriented in the space. This means that at macroscopic scale the heterogeneous media have no preferential directions even if the domains are anisotropic. The heterogeneous medium would behave as an isotropic solid.

Heterogeneity has a specific impact on elastic waves because these waves cross a great number of boundaries between adjacent domains which have different properties resulting in wave scattering, reflection, and refraction. Here, however, is an important point. Elastic waves can get reflected or transmitted only from objects that have bigger size than the wavelength. They get scattered from objects with equal or smaller size than the wavelength. If the dimensions of the randomly oriented in the space rock domains composing the heterogeneous Earth's crust have dimensions smaller than the wavelength of the seismic waves they cause scattering but have little impact on the wave propagation; the seismic waves do not 'feel' the presence of these domains and propagates as in an isotropic medium. If the domains' sizes are larger than the wavelength their boundaries have an impact on the waves' propagation. The waves get reflected and/or transmitted resulting in changes in polarization and velocity of propagation.

Rocks often have metal or metal oxides compositions. Metals are constructed by small single crystals that are oriented randomly in the space so their individual anisotropy does not affect the elastic waves that propagate in some direction. Metals and metal oxides of similar chemical composition form domains that have close to uniform distribution of density and elastic properties around the whole volume. Adjacent domains, however, could differ in density and elastic properties forming a heterogeneous structure.

The boundary conditions of the domains modify not only velocity of propagation and polarization but also the fronts of the elastic waves. Also the wave's polarization cannot be defined anymore as purely P- or S-type. The elastic waves in heterogeneous media of propagation are qP- and qS-waves. Medium's particles do not move on straight lines but follow complex trajectories that make the polarization a mixture of P- and S-modes which are not orthogonal to each other as it is in the case of a pure-mode plane wave.

In the hypocenter the bulk elastic wave may start its propagation as a longitudinal elastic wave. Through multiple reflections, refractions, and mode conversions at the boundaries between domains this P-wave generates a multitude of new P- and S-waves. Because of the complex boundary conditions the fronts of the waves get distorted to the point where they cannot anymore be classified as plane P-waves or S-waves. The elastic field gets more complex and all incident P-waves and S-waves reach the Earth's surface where under various angles of incidence laying in various planes. Many of these incident waves hit the surface under critical angles and form surface elastic waves some of which could have pretty well defined polarizations. The energy of the initial P-wave will be split among all new waves. All bulk P-, S-waves, and L-waves have the same frequency, but they all have different polarizations, wavelengths, velocities of propagation, and amplitudes. Because of the low attenuation in the metal oxide domains this multiplication process can continue longtime resulting in a great number of P- and S-waves propagating in all direction of the space randomly—they form an elastic field. In the next section we will analyze the process of the elastic mode conversion that contributes to the formation of specific surface seismic waves that prove to be the most destructive to buildings and cities.

1.4 Earthquake magnitude evaluation

Seismic waves with wavelengths of the order of dimensions of a building or shorter could cause worse damages if they hit such a building than waves with longer wavelengths because resonance effects. This can happen if the building gets in a cantilever resonance mode or standing elastic waves are created inside the building interior through reflections as shown in Fig. 1.4.1. On the left side of Fig. 1.4.1 cantilever oscillations are shown with their first, second, and third harmonics. On the right side a standing elastic wave is shown

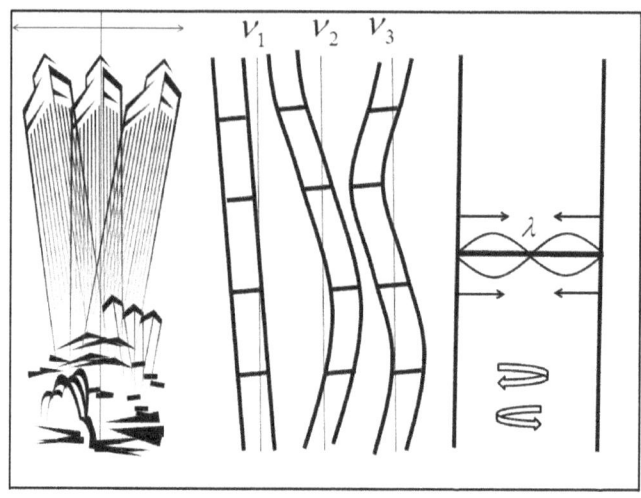

Fig. 1.4.1. Standing elastic waves cause resonance.

in its first harmonic. The standing wave created between the two reflecting walls is formed by superposition of a great number of reflected waves moving in opposite directions—from left to right and from right to left between the reflecting walls. All these waves have gone through different numbers of reflections from the walls but what is common to all of the waves that move in the same direction is that they will be all in phase if each wave changes its phase by $2\pi.n$, where an n is an integer number, at each reflection. When many in-phase waves move in the same direction a constructive interference occurs resulting in formation of a standing wave with larger amplitude than the amplitudes of the waves that form it. Separately twisting waves are generated as well rotating the building around its axis.

The cases of resonance shown in Fig. 1.4.1 can exit independently because they are caused by different oscillation modes. These are the simplest cases of resonance. There are many more possible resonance oscillations in the building that can occur when mechanical energy is supplied by an earthquake. For instance the building can also twist around its vertical axis or sidewinding. Being independent one from the other all these resonance conditions can exist individually at the same time. This means the building can oscillate as a cantilever, twist around is vertical axis, or be object of sidewinding motions. It is clear that all these modes of resonance oscillations can inflict construction damages. To avoid resonance oscillations during earthquakes buildings are designed in such a way that resonance conditions cannot occur in the usual frequency range of earthquakes.

To fight the destructive power of seismic waves it is necessary to measure accurately all parameters that characterize the earthquake. In 1935 Bruno Gutenberg and Charles Richter developed a scale that provides a relationship between the amplitude of oscillations measured by a seismometer at certain distance from the source of the earthquake and the magnitude of the earthquake. The seismometer measures the amplitude of the motion of the ground in microns (10^{-6} m). The Gutenberg-Richter magnitude scale is logarithmic which means that an earthquake of magnitude 5 is 10 times stronger than an earthquake of magnitude 4 and 100 time weaker than an earthquake of magnitude 7. The seismogram in Fig. 1.1.1.8 shows that the first wave that has arrived to the seismometer from a distant earthquake is a fast P-wave followed by a slower S-wave. The slowest of all are the surface waves or L-waves, but their amplitude is much greater than the amplitudes of the P- and S-waves.

The seismometer (called also seismograph) is a pretty simple machine. The principle of the detection of seismic waves has not evolved much since its invention in 132 AD in China. It represents a pendulum mass that moves when the ground shakes. The magnitude of the earthquake is estimated from the amplitude of the pendulum. Modern seismometers are more accurate since electronics and computers have been used to do the measurement. In modern seismometers the pendulum is kept immobile. What is measured is the force to keep it in its equilibrium position. During the earthquake tremor the force to retain the pendulum is proportional to the seismic wave's amplitude in the point where the seismometer is located. In most of the models this force is measured as an electrical signal which is digitized and processed by a computer. Modern seismometers measure the pendulum force in two perpendicularly to each other axes in the horizontal plane and in the vertical axis.

The only way to find out what type of seismic waves has taken part in a certain earthquake is the seismograms registered during the tremor. P-, S-, and L-waves are identified from the time of their arrival knowing that P-waves are the fastest ones, followed by S-waves, and finally by L-waves. The last portion of the seismogram representing a series of oscillations after the passage of the surface waves is summarized as S-coda waves or coda (Kanamori 1978). The name 'coda wave' also is used when referring to wave trains except direct waves. The waves at the tails of direct P- and S-waves are called P- or S-coda. Continuous wavetrains between P and S arrivals and those following S-arrival have not been explained so far. The direct S-wave amplitude decreases with increasing the distance from the epicenter, while S-coda wave amplitudes keep almost the same level for long time irrespectively of the distance from the epicenter (Sato 1997). An interesting observation has been reported about coda wave following the direct S-wave. The coda waves are composed by wavelets leaving the source region with the same slowness vector as the direct S-waves; however, later, S-coda waves are composed of wavelets leaving the source region in a variety of directions (Scherbaum et al. 1991). The transition between the two types takes place 1.5–2 times the S-wave travel time. It has also been observed that P-coda has almost the same propagation direction and velocity as the direct P-wave, while S-coda is composed by waves going in widely distributed directions of propagation and low semblance coefficient (Kuwahara et al. 1991). The data extracted from seismograms suggest the existence of heterogeneity in the Earth's lithosphere. Broadening of the enveloped of the S-wave seismograms of earthquakes recorded at long distances between 100 and 300 km has been observed (Sato 1997). The source duration of an earthquake having local magnitude less than 5 is shorter than 1s at close distances, while the duration of S-wave first arrival packets at long distances is much longer than 1s. This envelop broadening has been explained as due to strong diffraction and multiple forward scattering caused by slowly varying velocity structure in the heterogeneous lithosphere.

Table 1.4.1 summarizes the classes and magnitudes of according to Gutenberg and Richter scale. Earthquakes of number 9 or greater can cause ground topography changes resulting in devastating destruction to cities in the area.

The seismic wave amplitudes are so large that it is almost impossible for a normal man-made structure to survive, regardless of how good the building design was made or how strong the construction materials were. Such powerful earthquakes are rare. Gutenberg and Richter noticed that the lower the magnitude of an earthquake the higher the number of repeating earthquakes. Great earthquakes repeat at much lower frequency. This relationship between magnitude and frequency of repeating earthquakes has been called Gutenberg-Richter law.

From Table 1.4.1 we can see that there is some proportionality between the effect of destruction caused by an earthquake and its magnitude. However, earthquakes' statistics shows that there have been many exceptions so this proportionality cannot be assumed as being always a rule. It has happened that

Table 1.4.1. Gutenberg and Richter scale.

Class	Gutenberg-Richter	Effects	Number/year
great	>8	very destructive	0.1–0.2
major	7–7.9	serious damage	20
strong	6–6.9	a lot of damage in populated area	100
moderate	5–5.9	slight damage to buildings	500
light	4–4.9	felt with little damage	30,000
minor	3–3.9	felt with no damage	900,000

earthquakes of light or moderate magnitude have caused bigger destructions than strong or even major earthquakes. It looks like measuring the amplitude using the Gutenberg-Richter scale is not enough to fully characterize an earthquake regarding the effects of its destructive power. For this reason seismologists usually describe an earthquake using both Gutenberg-Richter scale and Mercalli scale.

In 1884, 50 years before Gutenberg-Richter scale, the Italian volcanologist Guiseppe Mercalli introduced a scale that measures the effects of an earthquake. Originally it has been a ten-degree scale. Later the scale has been expanded to twelve degrees and completely redesigned. If Gutenberg-Richter base-10 logarithmic scale measures the ground displacement caused by the earthquake using a seismograph, the Mercalli scale is established by observing the effects of the earthquake on earth's surface, objects, man-made structures, and humans (Table 1.4.1). Gutenberg-Richter scale is more technical than Mercalli scale but obviously it is not accurate enough to describe fully the parameters of an earthquake.

The modified Gutenberg-Richter scale is used to locate epicenters. This enables local building codes to establish standards for buildings which are

Table 1.4.2. Mercalli scale.

I. Instrumental	Detectable by seismometers, but not felt by people.
II. Weak	Felt by people only on upper floors of buildings.
III. Slight	Felt by people, but many do not recognize it as an earthquake. Vibrations similar to those of a passing truck.
IV. Moderate	Felt by all people. Some people wake up. Indoor objects shake. Cracking noise in the walls, but no damage.
V. Rather strong	Felt by people indoor and outdoor. Dishes and windows may break. Possible slight damage to buildings. Few people are frightened.
VI. Strong	Felt by all. Frightened people run outdoors. Windows dishes break, objects fall off the shelves. Some damage to poorly designed buildings.
VII. Very strong	Difficult to stand. Felt by people driving cars. Slight damage in buildings of good design, moderate in ordinary buildings, considerable damage in poorly designed buildings.
VIII. Destructive	Slight damage in buildings of good design, considerable damage in normal buildings, great damage in poorly designed buildings. Heavy damage in brick constructions. Heavy furniture move.
IX. Violent	People panic. Moderate damage in well-designed buildings, substantial damage in normal buildings with partial collapse.
X. Intense	Many well-done buildings destroyed. Most other structures destroyed with possibly shifted off foundations. Landslides.
XI. Extreme	Few buildings remain standing. Cracks and deformations of the ground.
XII. Catastrophic	Total destruction. Ground moves in waves. Large rocks move. Landscape altered. Rivers can change.

earthquake resistant. However a series of great earthquakes were poorly handled by the modified Richter scale such as the 1952 Aleutian Fox Islands quake and the 1960 Chilean quake (Aki 1972). The difficulties were explained with the size of these earthquakes that carried large amounts of energy. As a result, use of the modified Richter scale methodology, to estimate earthquake energy, was deficient at high energies (Kanamori 1978).

In 1972, Aki introduced elastic dislocation theory to improve understanding of the earthquake mechanism (Aki 1972). This theory proposed that the energy released from a quake is proportional to the surface area that breaks free, the average distance that the fault is displaced, and the rigidity of the material adjacent to the fault.

As it was pointed out earlier the type of a seismic wave and its polarization can have bigger impact to man-made structures than the magnitude of the earthquake. Neither Gutenberg-Richter nor Mercalli scale can characterize accurately an earthquake if used separately, however used together they complement each other. Farther in this book it will become clear why the magnitude of an earthquake is not necessarily the most important parameter and why the type of the elastic wave and its polarization could cause more damaging effects to buildings. Studying the properties of various types of bulk and surface elastic waves that can be generated and can propagate in areas with specific geological structures can help to answer the question what strategy would be more appropriate to fight the destructive power of a specific earthquake: to design buildings and use construction materials that are expected to withhold the seismic power or to design systems around cities that are capable to attenuate or deflect coming seismic waves. In certain geological systems it is possible to predict accurately the type of seismic waves and their characteristics that will occur during an eventual earthquake. In such cases it will be more efficient to design defensive systems around the cities capable of reflecting, diffusing, damping, or deflecting away seismic waves instead of relying solely on the strength of structures. This is a new and little explored field that could open the doors to a new world of opportunities for fighting the devastation consequences of earthquakes. Looking for solutions to deflect or dissipate seismic waves before they reach cities could change the whole concept of planning and designing man-made structures in areas where earthquakes often occur.

1.5 Linear elasticity theory

The main effect of an earthquake results in propagation of bulk longitudinal and shears seismic waves in the Earth's crust or mantle as well as surface seismic waves on the Earth's surface. In this chapter it will assume that all seismic waves—both bulk and surface—are elastic waves that linear and propagate in linear, isotropic, non-dispersive media. Usually seismic waves are nonlinear elastic waves and propagate in nonlinear, heterogeneous, anisotropic, dispersive media. We note that wave dispersion may be caused by attenuation and scattering of the waves in the medium of propagation (materials dispersion), and layered structures, waveguides, and free surfaces causing interference of the waves. Surface elastic waves are dispersive waves by their nature because usually they propagate on a half-space boundary surface. In many cases the linear approximation is pretty close to reality and can help to understand various phenomena. The way rocks deform is important for the analysis of seismic wave propagation. With respect to their elastic properties materials are (Nelson 2003): 1) brittle materials with large range of linear elasticity and small range of ductile behavior before fracture, and 2) ductile materials with small range of linear elasticity and large range of ductile elasticity before fracture (Fig. 1.5.1). Near the surface of the Earth behave in a brittle manner (Fig. 1.5.2.).

Crustal rocks are composed of minerals like quartz and feldspar which have high strength, particularly at low pressure and temperature. As we go deeper in the Earth the strength of these rocks initially increases. At a depth of about 15 km we reach a point called the brittle-ductile transition zone. Below this point rock strength decreases because fractures become closed and the temperature is higher, making the rocks behave in a ductile manner. At the base of the crust the rock type changes to peridotite which is rich in olivine. Olivine is stronger than the minerals that make up most crustal rocks, so the upper part of the mantle is again strong. But, just as in the crust, increasing temperature eventually predominates and at a depth of about 40 km the brittle-ductile transition zone in the mantle occurs. Below this point rocks behave in an increasingly ductile manner.

The crust of the Earth is composed of a great variety of igneous, metamorphic, and sedimentary rocks. The oceanic crust of the sheet is different from its continental crust. The oceanic crust is 5 km to 10 km thick and is composed primarily of basalt, diabase, and gabbro. The continental crust is typically from 30 km to 50 km thick and is mostly composed of slightly less dense rocks than those of the oceanic crust. Some of these less dense rocks, such as granite, are common in the continental crust but rare to absent in the oceanic crust. The seismic activities in the crust are usually the cause of earthquakes with high impact on the Earth's surface because the crust's brittle rocks a good medium of propagation for seismic waves. Brittle rocks form a low attenuation medium of propagation for longitudinal, shear, and surface elastic waves. The ductile rocks dissipate the elastic energy. Elastic waves cannot travel such long distances in ductile rocks as they do in brittle rocks.

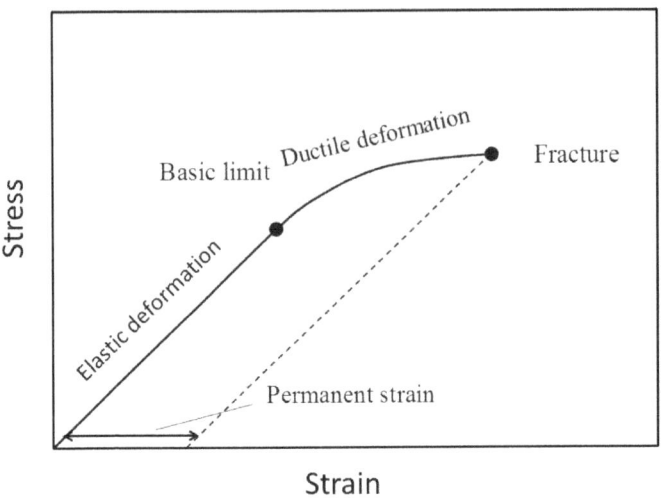

Fig. 1.5.1. Brittle and ductile stress-strain relationship.

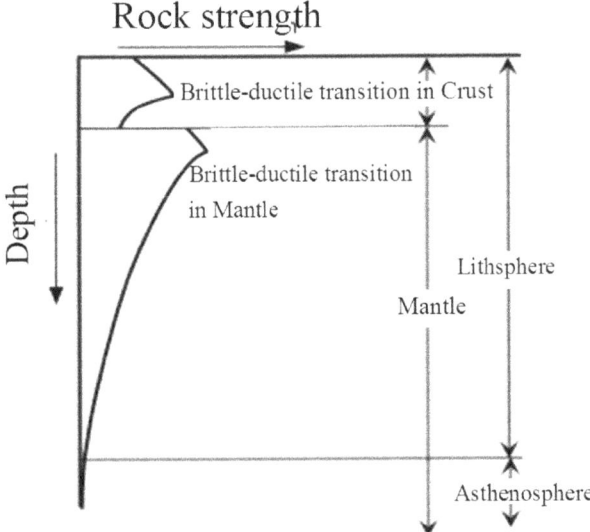

Fig. 1.5.2. Brittle and ductile rock.

1.5.1 Static elasticity. Strain-stress relationship

A seismic shock caused by some disturbance somewhere deep under the Earth's surface in the crust or mantle releases great amount of mechanical energy per unit of volume. The surrounding rocks are static, they do not move. Therefore the released mechanical energy cannot be kinetic energy; it is a potential energy. This potential energy is similar to the energy stored in compressed spring (Fig. 1.5.1.1). In this case the role of the spring is played by the rocks in the closest proximity to the focus. They are compressed, but the surrounding rocks retain them and do not let them to expand. Since the surrounding rocks are also immobile they will apply pressure (=force per unit of surface in N/m^2 or stress T) to the next layer of rocks. This pressure will propagate farther in the rock mass as a pressure wave. The process will last until the stored potential energy is totally dissipated in the rock mass and all rocks go back to equilibrium. The pulse lasts until the equilibrium is established. The potential energy is converted into kinetic energy and carried away by the pressure pulse that propagates across the rock mass as a pressure wave. Usually a series of aperiodic pulses—a pulse train—are generated in the hypocenter. They come one after the other at various intervals of time. Usually each pulse lasts in the range of milliseconds to seconds.

The rocks are not ideally rigid solids. The stress results not only in a pressure wave but also in a deformation or strain S. As all real solids the stress on the rocks results in some strain the level of which depends on where the rock scores between elasticity and plasticity. Elastic solid returns back to its initial shape when the stress stops acting. Ideally elasticity means that the solid will fully recover after the stress, while ideal plasticity means that the solid will remain deformed after the stress stops acting and will never recover to its initial shape. Of course there are various solids; some are more elastic than

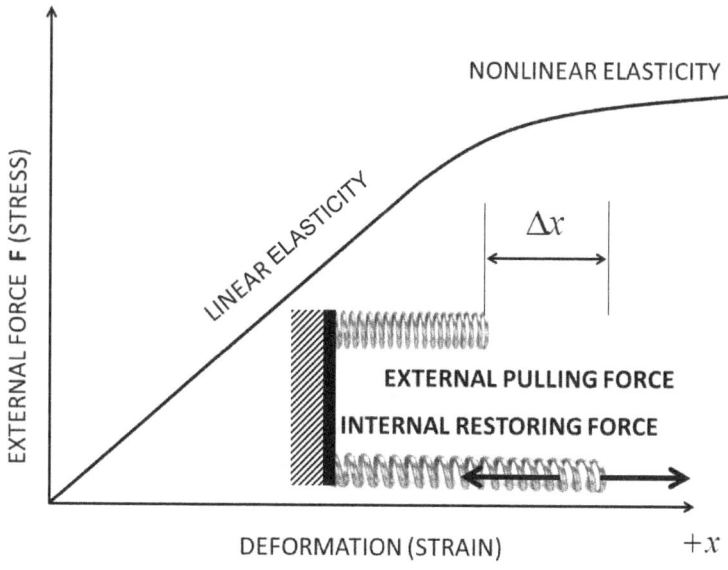

Fig. 1.5.1.1. Deformation of a spring. Hooke's law.

others. This means that they all get strained when stress is applied to them, but they go back to equilibrium in a different way. This strain is translated into a mechanical deformation (or displacement of particles) that form the rock. Since the particles are displaced from their equilibrium positions the rock reacts with an internal force that is trying to restore the equilibrium. An ideally rigid solid will not get deformed under stress. Therefore there would be no relationship between stress and strain. Real solids, however, get deformed and resist to the deformation trying to go back to equilibrium. If the stress—strain relationship is a straight line, which means that with increasing or decreasing stress strain follows a straight line without any hysteresis, Hooke's law rules in the range of linear elasticity only. In one dimension as the spring shown in Fig. 1.5.1.1 Hooke's law is $F = k\Delta x$, where F is the external force (stress) extending the spring and Δx is the deformation (strain) of the spring. The constant k is a positive number that characterizes the material of the spring.

Often Hooke's law is written in the form: $F = -k\Delta x$, where F is the internal force trying to restore the initial equilibrium status of the spring acting in the opposite direction.

In the case of three-dimensional elastic solid the external force applied to it will cause deformation. Depending on the elastic properties of the solid this will result into an internal reaction force trying to restore its initial equilibrium state (Fig. 1.5.1.2). The stress T_{ij} and the strain S_{kl} are tensors of second rank. The Hooke's law for a solid deformed by an external force is a linear stress-strain relationship:

$$T_{ij} = c_{ijkl}S_{kl} \tag{1.5.1.1}$$

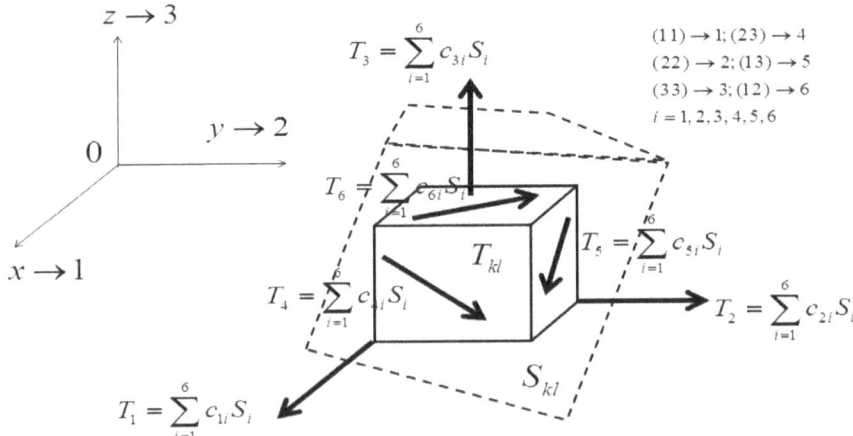

Fig. 1.5.1.2. External force (stress) causes a deformation (strain) S_{ij} expanding the solid.

In Eq. 1.5.1.1 the coefficients c_{ijkl} are called elastic stiffness coefficients. They form a tensor of rank four. We note that in the case of the one-dimensional spring the deformation is on the same axis X as the internal restoring force. In the case of a three-dimensional solid the deformation along the x-axis of the cube will result not only in a restoring force along the x-axis but also in all other directions (Fig. 1.5.1.2). The tensor equation Eq. 1.5.1.1 can be presented also in the following matrix form by replacing the indices $c_{ijkl} = c_{\alpha\beta} = c_{\beta\alpha}$ with α, $\beta = 1, 2, 3, 4, 5, 6$. We have $(\alpha, \beta) = (\beta, \alpha)$ because of the index symmetry. With $1 \rightarrow 11, 2 \rightarrow 22, 3 \rightarrow 33, 4 \rightarrow 23 = 32, 5 \rightarrow 13 = 31, 6 \rightarrow 12 = 21$ Eq. 1.5.1.1 can be rewritten in matrix form:

$$
\begin{pmatrix} T_1 \\ T_2 \\ T_3 \\ T_4 \\ T_5 \\ T_6 \end{pmatrix} = \begin{pmatrix} c_{11} & c_{12} & c_{13} & c_{14} & c_{15} & c_{16} \\ c_{21} & c_{22} & c_{23} & c_{24} & c_{25} & c_{26} \\ c_{31} & c_{32} & c_{33} & c_{34} & c_{35} & c_{36} \\ c_{41} & c_{42} & c_{43} & c_{44} & c_{45} & c_{46} \\ c_{51} & c_{52} & c_{53} & c_{54} & c_{55} & c_{56} \\ c_{61} & c_{62} & c_{63} & c_{64} & c_{65} & c_{66} \end{pmatrix} \begin{pmatrix} S_1 \\ S_2 \\ S_3 \\ S_4 \\ S_5 \\ S_6 \end{pmatrix}
\tag{1.5.1.2}
$$

Matrix multiplication yields to the stress components:

$$T_1 = c_{11}S_1 + c_{12}S_2 + c_{13}S_3 + c_{14}S_4 + c_{15}S_5 + c_{16}S_6$$

$$\cdots\cdots\cdots\cdots\cdots\cdots\cdots\cdots\cdots\cdots$$

$$T_6 = c_{61}S_1 + c_{62}S_2 + c_{63}S_3 + c_{64}S_4 + c_{65}S_5 + c_{66}S_6$$

The symmetry of the material will simplify the matrix. For example in the case of a material of cubic symmetry we have:

$$
\begin{pmatrix} c_{11} & c_{12} & c_{12} & 0 & 0 & 0 \\ c_{12} & c_{11} & c_{23} & 0 & 0 & 0 \\ c_{12} & c_{12} & c_{11} & 0 & 0 & 0 \\ 0 & 0 & 0 & c_{44} & 0 & 0 \\ 0 & 0 & 0 & 0 & c_{44} & 0 \\ 0 & 0 & 0 & 0 & 0 & c_{44} \end{pmatrix}
\tag{1.5.1.3}
$$

The matrix Eq. 1.5.1.3 for isotropic material is the same as for cubic crystal but in this case $c_{44} = (c_{11} - c_{12})/2$. In the case of a cubic crystal the matrix Eq. 1.5.1.3 is the same but this time c_{44} is not necessarily related to c_{11} and c_{12}. Each component of the stress can be obtained by multiplying the matrices of the compliance constants and the strain.

The tensor c_{ijkl} can be presented in the following form:

$$c_{ijkl} = \lambda \delta_{ij} \delta_{kl} + \mu_1 \delta_{ik} \delta_{jl} + \mu_2 \delta_{il} \delta_{jk} = \lambda \delta_{ij} \delta_{kl} + \mu \left(\delta_{ik} \delta_{jl} + \delta_{il} \delta_{jk} \right) \quad (1.5.1.4)$$

with $\mu = \mu_1 + \mu_2$ because of the symmetry of the tensor $(\alpha, \beta) = (\beta, \alpha)$. λ and μ are called Lamé constants and $\delta_{mn} = 0$ if $m \neq n$ and $\delta_{mn} = 1$ if $m = n$. The isotropic medium tensor Eq. 1.5.1.3 will have the following components:

$$c_{11} = c_{22} = c_{33} = \lambda + 2\mu$$

$$c_{12} = c_{23} = c_{13} = \lambda$$

$$c_{44} = c_{55} = c_{66} = \mu = \frac{c_{11} - c_{12}}{2}$$

$$(1.5.1.5)$$

Of course the stress cannot be increased infinitely. Each real solid has an upper limit of stress above which the stress-strain relationship is not linear anymore. At this point starts the range of ductile elasticity where the stress-strain relationship is not linear any more. Further increase of stress at the end of the ductile elasticity the solid will fracture (Beer 2009).

In the focus of the earthquake a stress-strain elastic field is created where the stress is the force applied per unit surface to the surrounding rocks. Depending of the seismic force and elastic properties of the rocks this will result in strain following the linear Hooke's law or the ductile curve. The rocks resist to the stressing force trying to return back to equilibrium. This situation is similar to the solid cube in Fig. 1.5.1.2. As a result of the stress the cube gets deformed (strained). Being an elastic solid if released it will stretch and bounce back to equilibrium retaking its initial shape. The stress-strain filed in the earthquake's focus causes the propagation of an elastic pulse wave with a semi-spherical front. As all solids the components of the Earth's crust and mantle have their ranges of linear elasticity, ductile elasticity, and fracture. The size of the focus of the earthquake and the elastic properties of the rocks will determine the parameters of the stress wave and its amplitude. Further propagation of the stress wave and its attenuation will be determined by the scattering in the heterogeneous structure of the rocks in the adjacent areas. Usually the stress wave is longitudinal elastic wave because the S-waves in heterogeneous media have much higher attenuation.

At the hypocenter a compressional pulse (or a series of compressional pulses) is generated by the seismic disturbance. The compressional pulse applies a stress on the surrounding rocks. The stress is translated into a mechanical deformation or displacement of particles of the rocks. Since the particles get displaced from their equilibrium positions the rock reacts with an internal force that is trying to restore the equilibrium. Figure 1.5.1.3 shows the deformation of the rock between two points A and B. The displacement between A and A' is $\vec{u}_A$ and between B and B' is $\vec{u}_B$ with

$$\vec{u}_A = \vec{u}_B + \vec{du} \text{ and } \vec{du} = \frac{\partial \vec{u}}{\partial x_i} dx_i.$$

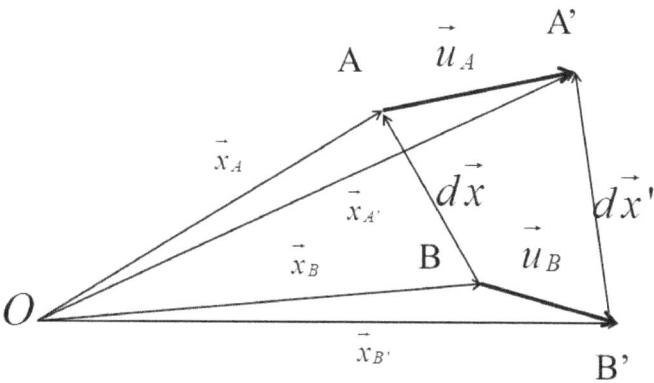

Fig. 1.5.1.3. Deformation of the medium of propagation.

The displacement resulting from the application of the force is

$$\left(d\vec{x}'\right)^2 = \left(d\vec{x}\right)^2 + 2\left(d\vec{x}\right).\left(d\vec{u}\right) + \left(d\vec{u}\right)^2$$

and also

$$\left(d\vec{x}'\right)^2 - \left(d\vec{x}\right)^2 = 2dx_i du_i + du_j du_j = 2\frac{\partial u_i}{\partial x_j}dx_i dx_j + \frac{\partial u_k}{\partial u_i}\frac{\partial u_k}{\partial u_j}dx_i dx_j.$$

Since $(i, j) = (j, i)$ which means that we get

$$\left(d\vec{x}'\right)^2 - \left(d\vec{x}\right)^2 = \left(\frac{\partial u_i}{\partial x_j} + \frac{\partial u_j}{\partial x_i} + \frac{\partial u_k}{\partial u_i}\frac{\partial u_k}{\partial u_j}\right)dx_i dx_j = 2S_{ij}dx_i dx_j$$

with:

$$S_{ij} = \frac{1}{2}\left(\frac{\partial u_i}{\partial x_j} + \frac{\partial u_j}{\partial x_i} + \frac{\partial u_k}{\partial u_i}\frac{\partial u_k}{\partial u_j}\right) \qquad (1.5.1.6)$$

If the deformation is small the quadratic term in Eq. 1.5.1.6 can be neglected because $\frac{\partial u_i}{\partial u_j} \ll 1$. In this case we have the linear elasticity strain relation:

$$S_{ij} = \frac{1}{2}\left(\frac{\partial u_i}{\partial x_j} + \frac{\partial u_j}{\partial x_i}\right) = S_{ji} \qquad (1.5.1.7)$$

The approximation $\frac{\partial u_i}{\partial u_j} \ll 1$ is linearizing Eq. 1.5.1.6 resulting to Eq. 1.5.1.7.

On the other hand the stress can be defined as the derivative of the compressional force applied in direction i to an element of surface a perpendicular to the axis k:

$$T_{ik} = \frac{dF_i}{da_k} \qquad (1.5.1.8)$$

An ideally rigid solid will not get deformed under strain; there is no relationship between strain and deformation. Real solids, however, get deformed and resist to the deformation trying to reach again equilibrium.

Since $c_{ijkl} = c_{klij}$ stress can be expressed as a function of deformation and we obtain the expression of Hooke's law (Eq. 1.5.1.1). We also have $c_{ijkl} = c_{jikl} = c_{ijlk} = c_{jilk}$. The stiffness constants have large values for rigid materials and small values for easily deformable solids.

Using Eq. 1.5.1.2 and Eq. 1.5.1.4 the stress-deformation relation can be expressed by:

$$T_{ij} = c_{ijkl} \frac{\partial u_i}{\partial x_k} \qquad (1.5.1.9)$$

The strain can be expressed as a linear function of the stress:

$$S_{ij} = s_{ijkl} T_{kl} \qquad (1.5.1.10)$$

In Eq. 1.5.1.10 the constants S_{ijkl} are called compliance constants. They measure the deformability of solid. They small values for rigid materials and large values for easily deformable solids. Stiffness is measured in N/m², as well as compliance because strain is dimensionless. Stiffness range is 10^9N/m² for soft materials like rubber and 10^{10}N/m² for rigid materials like crystals and rocks. Compliance range is 10^{-9} for rubberlike materials and 10^{-11} for rocks. Stress is in the range 10^7–10^8N/m² with strain $\sim 10^{-4}$–10^{-3}.

Typical characteristics of a propagating elastic wave are its amplitude A, frequency v, wavelength λ, phase velocity V, polarization (defining the wave's type or mode), and group velocity G (or energy flux, or Poynting vector). If the amplitude in the point $z = 0$ is A_0 the amplitude in the point z is $A = A_0 e^{-\alpha z}$, where α the attenuation coefficient of the elastic wave in dB per unit distance. We note that α is a function of frequency v, temperature T, and medium attenuation coefficient α'. In the Earth's crust around 1 Hz $\alpha \sim 10^{-3}$db/m.

2

Body Elastic Wave Propagation

In this Chapter we will discuss topics related to linear and nonlinear elastic wave propagation. Emphasis will be put on nonlinear wave propagation, because nonlinear phenomena are the most relevant to seismic waves and earthquakes. If the world were linear, there would be no earthquakes. Linear wave propagation will be discussed only as an approximation to the nonlinear wave processes.

2.1 Dynamic elasticity

The equation of wave propagation results from Newton's law: $\vec{F} = m\vec{a}$. Since the force density (=force par unit volume) with a stress T_{ij} is given by $f_i = \dfrac{\partial T_{ij}}{\partial x_j}$ the acceleration of a rock mass par unit of volume ρ (=density) is given by:

$$\rho \frac{\partial^2 u_i}{\partial t^2} = \frac{\partial T_{ij}}{\partial x_j} \qquad (2.1.1)$$

With the expression of the stress T_{ij} from Hooke's law we get the equation of wave propagation:

$$\rho \frac{\partial^2 u_i}{\partial t^2} = c_{ijkl} \frac{\partial^2 u_l}{\partial x_j \partial x_k} \qquad (2.1.2)$$

The solution of this equation for a plane wave propagating in a direction defined by $\vec{n}(n_1, n_2, n_3)$ is:

$$u_i = u_i^o F\left(t - \frac{\vec{n}.\vec{x}}{V}\right) = u_i^o F\left(t - \frac{n_j x_j}{V}\right) \qquad (2.1.3)$$

In this expression V is the wave phase velocity and u_i^o is the amplitude of displacement of the particles by the wave (polarization) which is parallel to the direction of propagation for a P-wave and perpendicular to this direction for an S-wave. By replacing u_i in the equation of propagation we get:

$$\rho V^2 u_i^o = c_{ijkl} n_j n_k u_l^o = \Gamma_{il} u_l^o \tag{2.1.4}$$

This equation is called Christoffel's equation.

$$\Gamma_{il} u_l^o = \rho V^2 u_i^o \tag{2.1.5}$$

The wave phase velocity and polarization can be found from the Christoffel's equation by calculating the eigenvector and eigenvalues of the secular equation:

$$\left| \Gamma_{il} - \rho V^2 \delta_{il} \right| = 0 \text{ with } \delta_{il} = 0 \text{ if } i \neq l \tag{2.1.6}$$

If a plane wave propagates in an isotropic continuum by using the Lamé coefficients:

$$\lambda + \mu = \frac{c_{11} + c_{12}}{2}, \quad \mu = \frac{c_{11} - c_{12}}{2} \tag{2.1.7}$$

Christoffel's propagation tensor becomes:

$$\Gamma_{il} = (\lambda + \mu) n_i n_l + \mu \delta_{il} n_k n_k \tag{2.1.8}$$

There are two possible solutions to this equation—one corresponding to a longitudinal wave with phase velocity $V_P = \sqrt{c_{11}/\rho}$ and one corresponding to a shear wave with velocity $V_S = \sqrt{(c_{11} - c_{12})/2\rho}$. It is clear that always the longitudinal wave is faster than the shear one because $V_S < V_P$. Since the solutions of the Christoffel's equation are the same for any direction of propagation in an isotropic medium of propagation we can conclude that these solutions are valid for any elastic wave in an isotropic medium, not necessarily just the plane ones. As we know this result is valid for seismic waves as well. The P-waves is always faster than the S-wave. This does not mean that we can make the conclusion that all seismic waves are linear elastic waves, nor that the Earth is an isotropic continuum. However, this result confirms that seismic waves are elastic waves. We will come to this discussion later in the book.

2.2 Reflection, refraction of elastic waves. Critical angles

In the previous section we demonstrated that in the earthquake focus the seismic shock generates bulk elastic waves in a certain frequency bandwidth.

These waves start their propagation in all directions initially with a semi-spherical front that gets progressively deformed as the waves cross various rock layers which form the heterogeneous structure of the Earth's before they reach the Earth's surface. At the boundaries between the rock domains the elastic waves undergo reflections, refractions, and mode conversions. Since the attenuation coefficients of metal-oxides rocks are very small a great number of elastic waves of various polarizations propagate in all directions forming diffuse elastic field. Such diffuse elastic fields exist almost permanently in the Earth's bulk without having any earthquake effects.

In this section we will see that elastic waves reflect and refract in a more complex way than light (electromagnetic waves) when crossing boundaries between media with different physical properties. If a plane-frontoptic wave crosses the boundary between two media with different optical properties (indexes of refraction) the reflection and refraction create new waves following Snell's law. Optic Snell's law is simpler—the only parameter that changes is the refraction angle as a function of indexes of refraction and angle of incidence. Elastic Snell's law is much more complicated and in certain cases only numerical computing can solve the problem of reflection and refraction. If the direction of propagation, polarization and amplitude of the incident wave, and the elastic properties of both media of propagation are known the problem of reflection and refraction consists in determining the directions of propagation, polarizations, amplitudes, and directions of group velocities. The starting point of the calculation is the boundary conditions between the two elastic media. If the two media are in contact then the boundary conditions are defined by 1) continuity of the displacement of particles at the interface, and 2) continuity of the elastic strain on each point of the boundary interface. In the simple case of free boundary of a solid in contact with air there is no refracted (transmitted) wave trough the boundary. There will be reflected waves only. The boundary condition in this case is very simple: the elastic strain is zero at each point of the interface boundary.

Consider two isotropic media 1 and 2 in contact. The boundary conditions are defined by 1) continuity of the displacement of particles at the interface, and 2) continuity of the elastic stress on each point of the boundary interface.

$$\vec{V_1} = \vec{V_2} \quad \vec{T_1} \cdot \vec{n} = \vec{T_2} \cdot \vec{n} \tag{2.2.1}$$

Since plane wave is described by $e^{-i\vec{k}\cdot\vec{r}} = e^{-i(k_x x + k_y y + k_z z)}$ the incident, reflected, and transmitted wave should have the same component of $\vec{k}$ parallel to the boundary. From $|\vec{k}| = \dfrac{2\pi}{\lambda} = \dfrac{2\pi\nu}{V} = \dfrac{\omega}{V}$ we can obtain Snell's law for an incident elastic P- or S-wave under an angle of incidence θ formed between the boundary and the vertical line defined by $\vec{n}$.

$$\frac{\omega}{V_1^P}\sin\theta_1^P = \frac{\omega}{V_1^S}\sin\theta_1^S = \frac{\omega}{V_2^P}\sin\theta_2^P = \frac{\omega}{V_2^S}\sin\theta_2^S \qquad (2.2.2)$$

In Eq. 2.2.2

$$\theta_1^P \equiv \theta_{1I}^P = \theta_{1R}^P;\, \theta_1^S \equiv \theta_{1I}^S = \theta_{1R}^S;\, \theta_2^S \equiv \theta_{2T}^S;\, \theta_2^P \equiv \theta_{2T}^P$$

Snell's law in electromagnetism is:

$$\frac{\omega}{V_1}\sin\theta_{1I} = \frac{\omega}{V_1}\sin\theta_{1R} = \frac{\omega}{V_2}\sin\theta_{2T} \qquad (2.2.3)$$

We can see that Eq. 2.2.3 is much simpler than Eq. 2.2.2.

In the simple case of a free boundary of a solid in contact with a fluid there is no transmitted (refracted) elastic wave trough the boundary. There will be reflected waves only. The boundary condition in this case is very simple: the elastic strain is zero at each point of the interface boundary $\vec{T_1} \cdot \vec{n} = 0$.

Figure 2.2.1a shows the simplest case of a SH-wave polarized | | to the boundary. Figure 2.2.1.b and Figure 2.2.1.c show the reflection and refraction of an incident P-wave and an incident SV-wave (the vertical polarization is in the plane $\perp$ to the boundary, called sagittal plane). In both cases two reflected and transmitted waves—one P-type and one SV-type—are created. The P-waves are always faster than the S-waves. The intersection point of all reflected and refracted waves of a specific mode form a closed curve in the space called slowness surface.

Figure 2.2.2 shows the slowness surfaces on a boundary between isotropic fused quartz and anisotropic α-quartz (α-SiO$_2$). The slowness surfaces of anisotropic solid such as α-quartz are usually three—one for the P-wave, one for the faster S$_1$-wave and one for the slower S$_2$-wave. Both S-waves have perpendicular polarizations and equal velocities of propagation (degenerated S-waves) in certain directions called pure-mode axes. In the general case the S$_1$- and S$_2$-waves are quasi shear waves and propagate with different velocities. At a boundary with an isotropic solid as shown in Fig. 2.2.2 their refracted components become degenerated waves. The slowness surfaces of isotropic solids are always two spheres, one with bigger radius (S-wave) than the slowness sphere of the P-wave. The intersection point of the slowness surfaces for P-waves and S-waves are shown in Fig. 2.2.1 (the vertical dashed line). Since P-waves are faster than S-waves their slowness surface will be enclosed in the slowness surface of S-waves. Their angles of reflection and transmissions are bigger than the angles of reflections of S-waves. We remind that the two media in Fig. 2.2.1 are isotropic. In some cases of anisotropic media the refraction is more complex and it not always possible to determine the slowness surfaces as a set of intersection points.

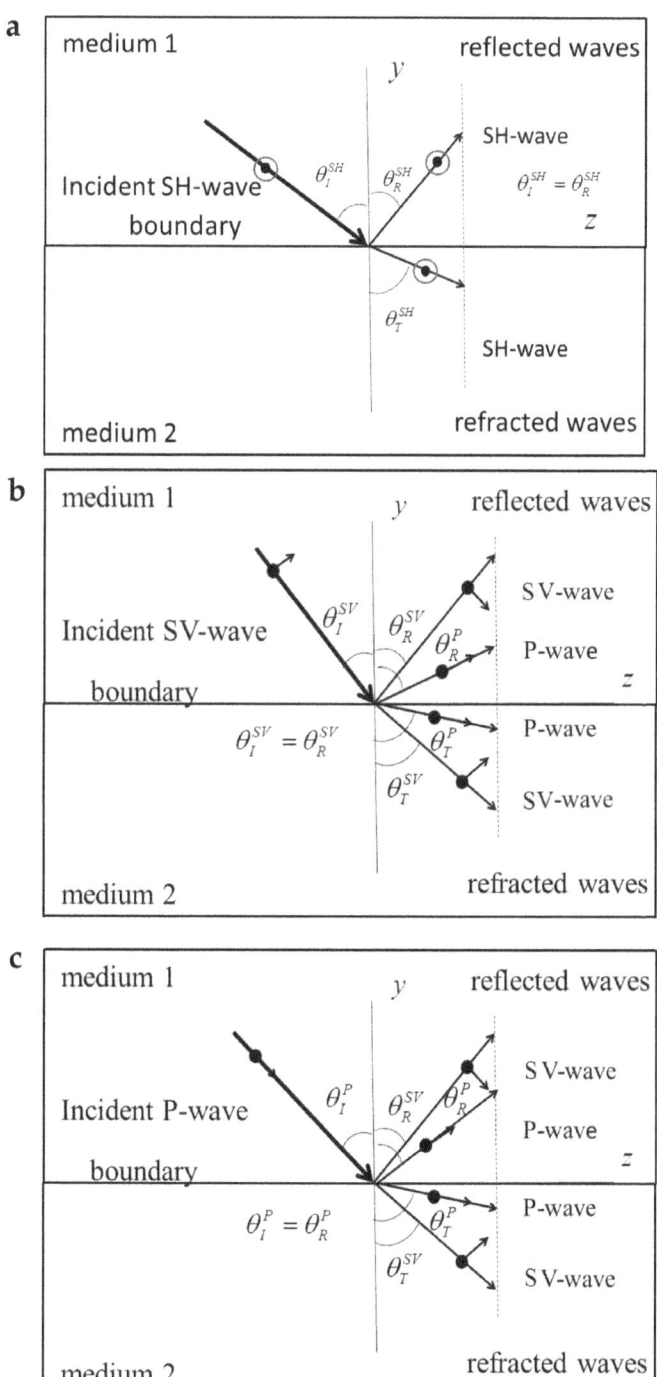

Fig. 2.2.1. (a) SH-wave incidence; (b) SV-wave incidence; (c) P-wave incidence.

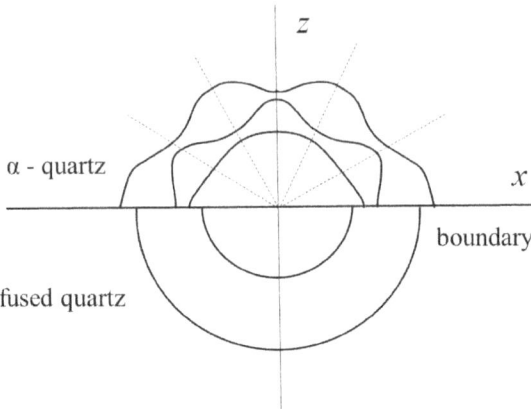

Fig. 2.2.2. Slowness surfaces on a boundary between isotropic fused quartz and anisotropic quartz.

2.2.1 SH-, SV-, and P-wave incidence

The simplest case of refraction is when a horizontally polarized (the displacement of the particles is in the plane that is parallel to the boundary between media 1 and 2) shear wave—a SH-wave (Fig. 2.2.1a). In this case the polarization does not have any component on the axis perpendicular to the boundary. The energy of the incident SH-wave will be split between a reflected SH-wave propagation back into medium 1 under the same angle as the incident HS-wave and a transmitted SH-wave which angle will depend on the ratio of the velocities in medium 1 and medium 2.

From the boundary conditions (Eq. 2.2.1) it follows that an incident to the boundary between the media 1 and 2 shear wave with a horizontal polarization (displacement of particles is parallel to the boundary) the reflected and transmitted waves will have the same polarization. All refracted waves are SH-waves because there is no polarization component on the vertical line defined by $\vec{n}$. In this case the Snell's law is:

$$\frac{\sin \theta_I^{SH}}{\sin \theta_T^{SH}} = \frac{V_1}{V_2} \tag{2.2.1.1}$$

The reflection and transmission coefficients are functions of elastic impedances of the two media:

$$R_{SH}^{SH} = \frac{Z_1 \cos \theta_I^{SH} - Z_2 \cos \theta_T^{SH}}{Z_1 \cos \theta_I^{SH} + Z_2 \cos \theta_T^{SH}}$$

$$T_{SH}^{SH} = \frac{2Z_1 \cos \theta_I^{SH}}{Z_1 \cos \theta_I^{SH} + Z_2 \cos \theta_T^{SH}} \tag{2.2.1.2}$$

For an elastic wave falling normally to the boundary we get:

$$R^{SH} = \frac{Z_1 - Z_2}{Z_1 + Z_2} = \frac{Z_1 / Z_2 - 1}{Z_1 / Z_2 + 1}$$

$$T = 1 - R = \frac{2Z_1}{Z_1 + Z_2} = \frac{2Z_1 / Z_2}{Z_1 / Z_2 + 1} \qquad (2.2.1.3)$$

If $Z_2/Z_1 < 1$ the reflection coefficient will go to zero for an angle of incidence

$$\theta^{SH}_{I_{critical}} = Arc \cos \left(\frac{Z_2}{Z_1} \cos \theta^{SH}_T \right)$$ and no reflection back into medium 1 will take

place—there will be total transmission into medium 2. If $V^{SH}_2 / V^{SH}_1 > 1$ for all angles of incidence $\theta^{SH}_I > \theta^{SH}_{I_{critical}}$ a total reflection back into medium 1 occurs with an evanescent transmitted HS-wave into medium 2. In this case the reflection coefficient goes to 1.

In oblique incidence the reflection and transmission coefficients depend on the angle of incidence, therefore on the velocity ratio V_2/V_1 according to Eq. 2.2.1.1, and the impedance ratio according to Eq. 2.1.2.2 of the media 1 and 2. If $V_2/V_1 < 1$ then $\theta^{SH}_T > \theta^{SH}_I$ and if $Z_2/Z_1 < 1$ there is no critical angle for the transmitted wave and for some angle of incidence R goes to 0. If $Z_2/Z_1 > 1$ the reflection coefficient never becomes 0. For $V_2/V_1 > 1$ there is a critical angle where total reflection occurs. The wave is reflected entirely back in medium 1 while the transmitted wave's amplitude in medium 2 is evanescent. R becomes 0 only when $Z_2/Z_1 > 1$ because in this case $\theta^{SH}_T > \theta^{SH}_I$.

The coefficients of reflection and transmission of the elastic SH-waves depend on the angle of incidence and the impedance and velocity ratios of the two media 1 and 2. In the case of $V^{SH}_2 / V^{SH}_1 < 1$ we have $\theta^{SH}_T / \theta^{SH}_I < 1$ and therefore there is no critical angle. If $Z_2/Z_1 < 1$ the reflection coefficient will go to zero for an angle of incidence $\theta^{SH}_{I_{critical}}$ and no reflection back into medium 1 will take place—there will be total transmission into medium 2. If $V^{SH}_2 / V^{SH}_1 < 1$ for all angles of incidence $\theta^{SH}_I > \theta^{SH}_{I_{critical}}$ a total reflection back into medium 1 occurs with an evanescent transmitted SH-wave into medium 2. In this case the reflection coefficient goes to 1.

The SH-wave reflection and refraction (transmission) only SH-waves are involved—one reflected and one refracted (transmitted). There is no P-wave. The case of SV-wave (vertically polarized) as shown in Fig. 2.2.1b is different—both SV- and P-waves are excited as reflected and transmitted waves. The detailed calculation method of the reflection and transmission coefficients can be found in Auld (Auld 1973). The calculation of critical angles and exact reflection and transmission coefficient requires numerical computation. Similar is the case of an incident P-wave.

2.2.2 Free solid-fluid boundary

In the case of a free boundary a solid is in contact with a fluid—air, water, etc. There are only reflected waves in the solid. No waves are transmitted into the fluid. P-waves are allowed to propagate in a fluid. S-waves are prohibited. However the elastic impedance of a solid is much higher than the elastic impedance of a fluid, so if a P-wave is incident to the free boundary the reflection coefficient will be close to 1 because $Z_2/Z_1 << 1$. The boundary condition for an incident wave is simple: the perpendicular to the free surface component of the elastic strain on each point of the boundary interface is zero (Fig. 2.2.2.1a) reflected from the free boundary. By contrast, an incident SV-wave is not totally reflected. Because of mode conversion it is reflected into two waves—one the same and the other of the other type Fig. 2.2.2.1a.

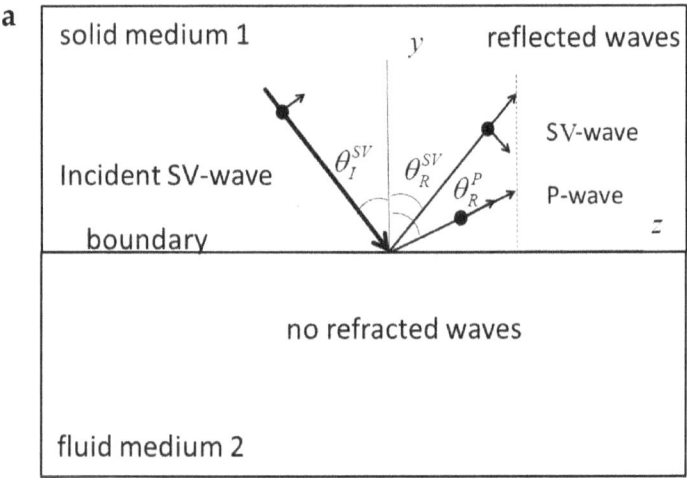

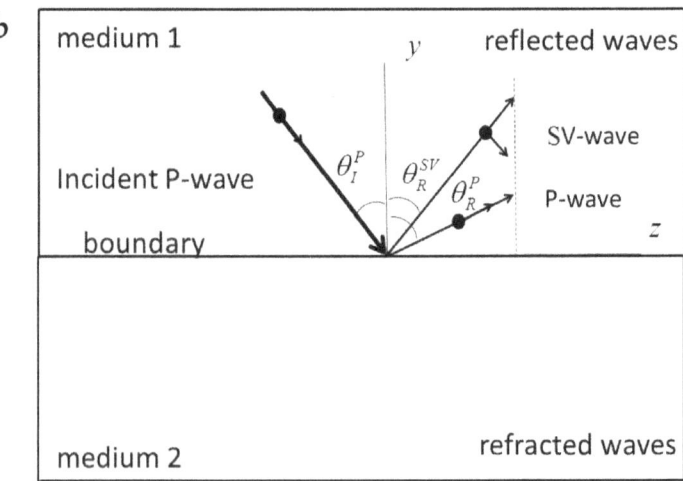

Fig. 2.2.2.1. (a) SV-wave incidence to a half space free boundary; (b) P-wave incidence to a half space free boundary.

Figure 2.2.2.1b shows an incident longitudinal wave to the boundary between isotropic medium 1 and fluid medium 2. There are only reflected waves—one P-wave propagating under the same angle as the angle of incidence and one SV-wave polarized vertically in the plane that is perpendicular to the boundary. This vertical plane also contains the incident P-wave. The SV-wave propagates under a smaller angle than the P-wave. Since the angle of incidence is equal to the angle of reflection for the P-wave the angle of reflection of the SV-wave is defined by Snell's law.

$$\frac{\sin \theta_P^P}{\sin \theta_{SV}^{SV}} = \frac{V_P}{V_S}$$

The reflection coefficients for an incident SV-wave are:

$$R_P^{SV} = \frac{2\left(V_P / V_S\right)\sin 2\theta_I^{SV} \cos 2\theta_I^{SV}}{\sin 2\theta_I^{SV} \sin \theta_R^P + \left(V_P / V_S\right)^2 \cos^2 2\theta_I^{SV}}$$

$$R_{SV}^{SV} = -\frac{\sin 2\theta_I^{SV} \sin 2\theta_R^P - \left(V_P / V_S\right)^2 \cos^2 2\theta_I^{SV}}{\sin 2\theta_I^{SV} \sin \theta_R^P + \left(V_P / V_S\right)^2 \cos^2 2\theta_I^{SV}}$$

(2.2.2.1)

The coefficient of reflection of the SV-wave is positive for any angle of incidence. From Fig. 2.2.2.1b it can be seen that there is a critical angle $\theta_{Icritical}^P$ beyond which only the SV-wave is reflected. The incident P-wave is converted entirely into an S-wave with vertical polarization—SV-wave.

The reflection coefficients for an incident P-wave are:

$$R_P^P = \frac{\sin 2\theta_R^{SV} \sin 2\theta_I^P - \left(V_P / V_S\right)^2 \cos^2 2\theta_R^{SV}}{\sin 2\theta_R^{SV} \sin 2\theta_I^P + \left(V_P / V_S\right)^2 \cos^2 2\theta_R^{SV}}$$

$$R_{SV}^P = \frac{2\left(V_P / V_S\right)\sin 2\theta_I^P \cos 2\theta_R^{SV}}{\sin 2\theta_R^{SV} \sin 2\theta_I^P + \left(V_P / V_S\right)^2 \cos^2 2\theta_R^{SV}}$$

(2.2.2.2)

We note that in Eq. 2.2.2.1 and Eq. 2.2.2.2 $\theta_I^{SV} = \theta_R^{SV}$ and $\theta_I^P = \theta_R^P$.

There are no critical angles for an incident P-wave. In both cases of SV- and P-wave incidence there are two angles of incidence at which the incident wave is totally reflected into the other type of wave—P-wave into SV-wave and SV-wave into P-wave.

Equation 2.2.2.1 and Eq. 2.2.2.2 yield to:

$$R_P^P = -R_{SV}^{SV}$$ (2.2.2.3)

$$\left(R_P^P\right)^2 + R_P^{SV} R_{SV}^P = 1$$ (2.2.2.4)

2.2.3 Free isotropic plate

Let consider a plate with two free boundary conditions with a SH-wave propagating between them as shown in Fig. 2.2.3.1a,b,c (Auld 1973). The plate forms an acoustic waveguide. In order to keep on propagating all SH-waves going up and down should obey the transverse resonance principle, which means that their phase should shift of n-times of 2π every round trip where n is an integer. In this case all waves going up and all waves going down will be in phase and could interfere in a constructive way maintaining the propagation in the waveguide. If the condition of a phase shift equal to $2n\pi$ is not valid the SH-waves will become quickly evanescent and there will be no farther propagation. The wave mode will have even symmetry for even n and odd symmetry for n odd. If the plate boundaries are loss free the amplitudes of the waves going up and those going down differ by the sign $+/-$ only with a reflection coefficient equal to 1 because we deal with horizontally polarized waves (polarization along the X axis only) and no transmitted waves are allowed.

From Fig. 2.2.3.1b we can see that if the wave's frequency increases the angle θ increases as well because we know that $V = \lambda \upsilon = \lambda \omega / 2\pi = (2\pi/k)(\omega/2\pi) = k/\omega$ or $1/V = k/\omega$, where $|\bar{k}|$ is the wave vector module.

The relation between velocity of propagation and frequency is called dispersion and the wave is called respectively a dispersive wave. For decreasing frequency the incidence angle θ decreases and becomes zero for some frequency value that is called cutoff frequency—$\omega_{cutoff} = (n\pi/b)V_S^H$. This means that for all $\omega > \omega_{cutoff}$ we have propagation in the waveguide and for all frequencies $\omega < \omega_{cutoff}$ the wave becomes exponentially evanescent away from its source of generation and there is not acoustic energy transport along the plate. Each SH-mode that can propagate without losses along the plate has a specific value of the integer n and has its own dispersive curve shown in Fig. 2.2.3.1c.

The case of a SH-wave propagating in the free boundary plate is the simplest one. Other modes can also propagate along the plate such as SV-wave and P-wave. However the SV- and P-waves cannot propagate individually; they are coupled. Since the P- and SV-waves propagates at different velocities (P-waves are faster than SV-waves) but also they must have the same component of their wave vector on the axis of propagation Z their angles of incidence are different. The angles of incidence of the P-mode is bigger that the angle of incidence of the SV-mode. The propagation of these coupled waves called Lamb waves is shown in Fig. 2.2.3.2.

Let consider Fig. 2.2.3.2 again following the transverse resonance analysis (Auld 1978). The incident waves are $A_P e^{-i\bar{k}_P \cdot \bar{r}}$ and $A_S e^{-i\bar{k}_S \cdot \bar{r}}$. Since the propagation is along z-axis the wave number $k_z = \beta$ of both P- and S-wave is the same.

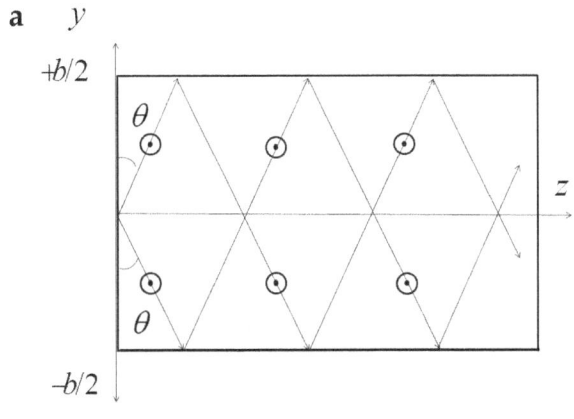

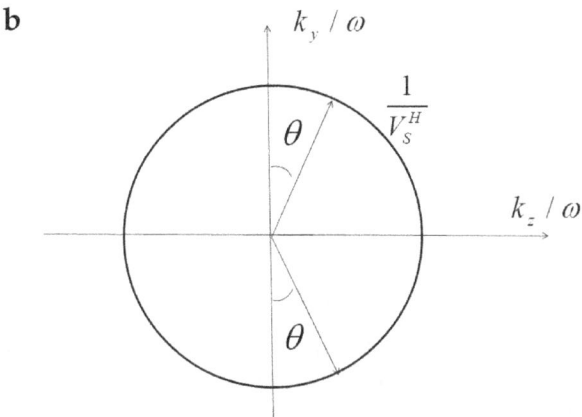

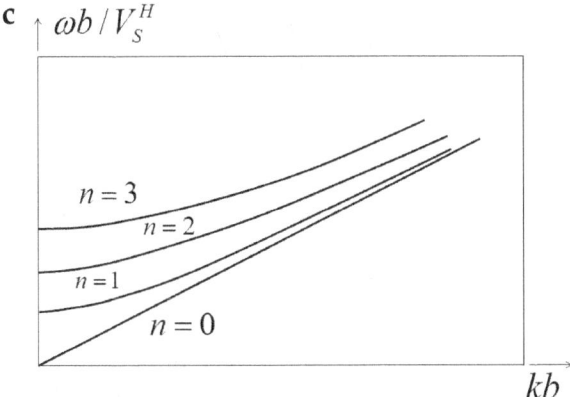

Fig. 2.2.3.1. (a) SH-wave propagating in a plate; (b) If the wave's frequency increases the angle θ increases; (c) Dispersive curves.

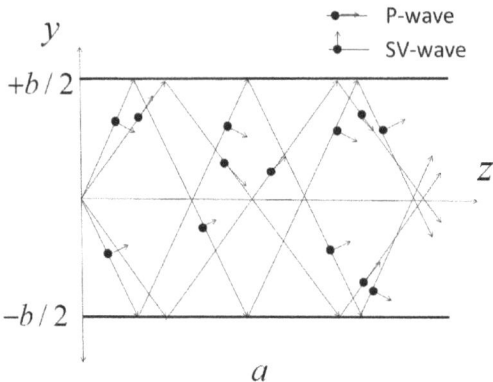

Fig. 2.2.3.2. Lamb waves are coupled P-modes and SV-modes.

The reflected waves are respectively $B_p e^{-i\vec{k}_p \cdot \vec{r}}$ and $B_S e^{-i\vec{k}_S \cdot \vec{r}}$. Since we consider the reflections from the wall without energy leaks we have $B_p = \pm A_p$ and $B_S = \pm A_S$. Taking into account the mode conversion during the reflection at the boundary $y = -b/2$ we have:

$$\pm A_p e^{ik_{tP}b/2} = R_{RR} A_p e^{-ik_{tP}b/2} + R_{RS} A_S e^{-ik_{tS}b/2}$$
$$\pm A_S e^{ik_{tS}b/2} = R_{SR} A_p e^{-ik_{tP}b/2} + R_{SS} A_S e^{-ik_{tS}b/2}$$

$$(2.2.3.1)$$

In Eq. 2.2.3.1 k_{tP} and k_{tS} are the transverse wave vector components of the P- and S-waves. The mode are either symmetric or antisymmetric depending number of reflections.

Solving Eq. 2.2.3.1 yields to the dispersion Rayleigh-Lamb relations:

$$k_{tP}^2 = \left(\frac{\omega}{V_P}\right)^2 - \beta^2$$

$$(2.2.3.2)$$

$$k_{tS}^2 = \left(\frac{\omega}{V_S}\right)^2 - \beta^2$$

The Lamb waves are guided and strongly coupled waves, i.e., they can propagate only together. The P- and SV-modes shift to each other for both symmetric and antisymmetric modes. The dispersion curves for the fundamental modes are presented in Fig. 2.2.3.3.

The wave coupling phenomenon is very important in seismic waves. The propagation of Lamb waves has much more complicated characteristics than SH-modes with 3-dimensional dispersion curves. We will return to Lamb waves in Section 3.2 of this book to introduce other types of surface elastic waves following the elastic waveguide theory. Lamb waves have many

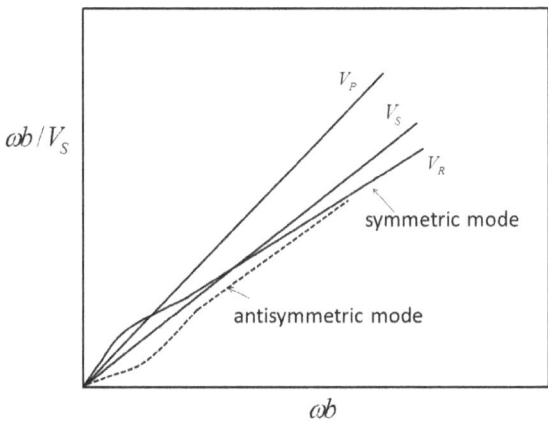

Fig. 2.2.3.3. Dispersion curves for the fundamental modes.

applications in physical acoustics in various acoustoelectronics devices where acoustic waveguides are needed. Since they can propagate in free boundary plates only they are not relevant to seismology and we will not go any further in a detailed study. However, as we will see in the next section, Lamb waves can help us to understand the nature of other type of surface acoustic waves called Rayleigh waves that have a big impact in the study of seismic waves.

2.3 Elastic wave energy transport. Poynting vector

Besides on the moment magnitude of an earthquake its effects depend also on the type and polarization of the elastic waves that have been generated not only at the earthquake's focus but also at the Earth's surface as a result of elastic mode conversions. A propagating elastic wave always transports mechanical energy. During various mode conversions this energy is conserved. This means all new elastic waves created should 'inherit' part of the energy initially released at the hypocenter. Since the elastic waves propagate in various directions covering huge volume or surface (depending on whether they are bulk or surface waves) it would be more appropriate to consider the energy per unit volume or unit surface called energy density. The energy density at various points around the epicenter can vary on a large scale depending on the elastic energy transport and the type of elastic waves carrying this energy. The geophysical specifications of the region have a significant impact on the elastic wave propagation and the energy transport.

The Poynting vector represents the rate of energy transfer per unit area in units of watts per square meter W/m^2. The flux of Poynting vector is defined by Poynting theorem. The energy transport can be presented as Poynting vector.

The energy density is the energy contained in unity of volume. The total energy of some volume will be the sum (or integral) of the energy density over this volume. If the energy is transported in the space as it happens when an elastic wave propagates through some medium then we talk about energy flux which is the Poynting vector (=rate of energy transfer per unit area) integrated over the surface crossed. In vector calculus, divergence is a vector operator that measures the magnitude of a vector field's source or sink at a given point, in terms of a signed scalar. More technically, the divergence represents the volume density of the outwardflux of a vector field from an infinitesimal volume around a given point as shown in Fig. 2.3.1.

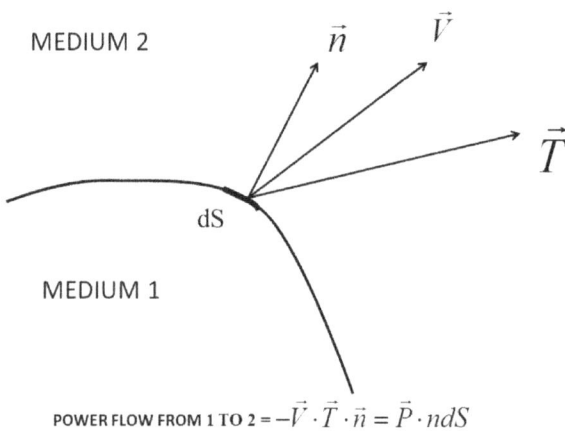

POWER FLOW FROM 1 TO 2 $= -\vec{V} \cdot \vec{T} \cdot \vec{n} = \vec{P} \cdot n dS$

Fig. 2.3.1. Power flow, phase velocity and Poynting vector.

Poynting vector has been used initially in electromagnetism $(\vec{P} = \vec{E} \times \vec{H}$ where $\vec{E}$ is the electric field and $\vec{H}$ is the magnetic energy) to describe the electromagnetic energy flux vector, but since any type of energy transport is characterized by its direction of movement in the space as well as density later Poynting vector has been used to describe also mechanical energy transport.

In the propagation of an elastic wave the energy enclosed in the volume **v** (Fig. 2.3.2) surrounded by surface **s** will vary as a function of time. The total energy carried by an elastic wave is defined by (Dieulesaint and Royer 1974):

$$\delta E_T = \delta E_k + \delta E_p = \int_v \rho \frac{\partial^2 u_i}{\partial t} du_i dv + \int_v T_{ik} dS_{ik} dv \qquad (2.3.1)$$

Since $du_i = \frac{\partial u_i}{\partial t} dt$ the kinetic energy becomes:

$$\delta E_k = \int_v \frac{\partial \Sigma_k}{\partial t} dt dv$$

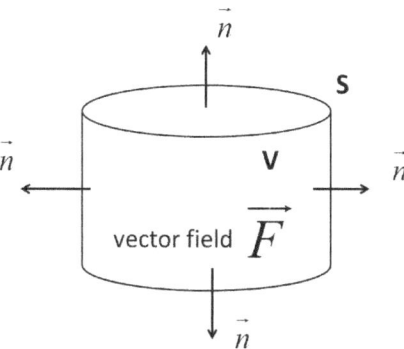

Fig. 2.3.2. Poynting vector.

The kinetic energy density Σ_k is equal to:

$$\Sigma_k = \frac{1}{2}\rho\left(\frac{\partial u_i}{\partial t}\right)^2 \qquad (2.3.2)$$

If the variation of the potential energy is $d\Phi = T_{ik}dS_{ik}$ the total energy in the volume **v** equal to:

$$E_T = \int_v d\left(\Sigma_k + \Phi_p\right)dv \qquad (2.3.3)$$

During the propagation of the elastic wave the energy E in a volume **v** varies by $dE = \int_S T_i\left(\vec{l}\right)du_i ds = \int_S T_{ik}l_k du_i ds$ where $T(\vec{l})$ applied in each point of the surface **s** as a function of time t:

$$\frac{dE}{dt} = \int_S T_{ik}\frac{\partial u_i}{\partial t}l_k ds \text{ or } \frac{dE}{dt} + \int_S P_k l_k ds = 0 \qquad (2.3.4)$$

The vector $\vec{P}$ is the elastic Poynting vector:

$$P_k = -T_{ik}\frac{\partial u_i}{\partial t} \qquad (2.3.5)$$

Equation 2.3.4 shows that the variation of the total energy in a volume **v** is due to the flux of the vector $\vec{P}$ across the surface **s** as shown in Fig. 2.3.2. If $\vec{V}^E$ is the velocity of the energy transport (or group velocity) and Σ is the total energy density the Poynting vector is

$$\vec{P} = \Sigma\vec{V}^E \qquad (2.3.6)$$

In the case of an anisotropic medium of propagation the velocity of the energy transport and the velocity of propagation of a plane elastic wave do not have the same direction and scalar value (see Fig. 2.3.1).

The minus sign can be explained using Fig. 2.3.1. The power delivered to an object is the force applied multiplied by the velocity. The force applied through the surface dS from medium 1 to medium 2 is equal to $-\vec{T} \cdot \vec{n} dS$ and, therefore the power through dS from medium 1 to medium 2 is $-\vec{V} \cdot \vec{T} \cdot \vec{n} dS$. $\vec{V}$ is in m/sec, $\vec{T}$ is in newton/m² and $\vec{P}$ is newton/m.sec or watt/m². The Poynting vector represents the power density.

2.4 Attenuation of elastic waves in isotropic solids

Consider again Fig. 2.3.2 where a solid with a volume $\delta \mathbf{v}$ surrounded by a surface $\delta \mathbf{s}$ is moving under the action of a force $\vec{F}$, called traction force. The forces associated with the motion of the solid are the body force $\vec{F} \delta \mathbf{v}$ and the traction forces applied to its surface $\delta \mathbf{s}$ the neighboring solids. The applied force is $\vec{T} \cdot \vec{n}$ acting on the surface, i.e.,

$$\int_{\delta S} \vec{T} \cdot \vec{n} dS$$

Newton's law gives:

$$\int_{\delta s} \vec{T} \cdot \vec{n} ds + \int_{\delta v} \vec{F} dv = \int_{\delta v} \rho \frac{\partial^2 \vec{u}}{\partial t^2} dv$$

If the volume of the solid is sufficiently small the equation of motion Eq. 2.2.1 becomes:

$$\vec{\nabla} \cdot \vec{T} = \rho \frac{\partial^2 \vec{u}}{\partial t^2} - \vec{F} \tag{2.4.1}$$

Equation 2.1.1 describes the static elasticity of the solid, while Eq. 2.4.1 is more appropriate for a propagating elastic wave displacing particles in the medium.

Since

$$\vec{V} = \frac{\partial \vec{u}}{\partial t} \text{ and } \vec{p} = \rho \vec{V}$$

where $\vec{p}$ is the momentum density, Eq. 2.4.1 becomes:

$$\vec{\nabla} \cdot \vec{T} = \rho \frac{\partial \vec{V}}{\partial t} - \vec{F} = \frac{\partial \vec{p}}{\partial t} - \vec{F} \tag{2.4.2}$$

In a source-free field $\vec{F} = 0$ and a plane wave propagating in direction $\vec{n}$ = $n_1\vec{x} + n_2\vec{y} + n_3\vec{z}$ is proportional to $e^{i(\omega t - \vec{k}\cdot\vec{n})}$. Equation 2.4.2 can be rewritten in the form:

$$\vec{\nabla}\cdot\frac{\partial\vec{T}}{\partial t} = \rho\frac{\partial^2\vec{V}}{\partial t^2} - \frac{\partial\vec{F}}{\partial t}$$

or

$$\nabla_{i\alpha}c_{\alpha\beta}\nabla_{\beta j}V_j = \rho\frac{\partial^2 V_i}{\partial t^2} - \frac{\partial F_i}{\partial t} \tag{2.4.3}$$

The operators $\nabla_{i\alpha}$ and $\nabla_{\beta j}$ with $i, j = 1,2,3$ and $\alpha, \beta = 1,2,3,4,5,6$ are defined by (Auld 1973):

$$\vec{\nabla}\cdot \rightarrow \nabla_{i\alpha} = \begin{bmatrix} \dfrac{\partial}{\partial x} & & & & \dfrac{\partial}{\partial z} & \dfrac{\partial}{\partial y} \\ & \dfrac{\partial}{\partial y} & & \dfrac{\partial}{\partial z} & & \dfrac{\partial}{\partial x} \\ & & \dfrac{\partial}{\partial z} & \dfrac{\partial}{\partial y} & \dfrac{\partial}{\partial x} & \end{bmatrix} \tag{2.4.4}$$

$$\nabla_s \rightarrow \nabla_{\beta j} = \begin{bmatrix} \dfrac{\partial}{\partial x} & & \\ & \dfrac{\partial}{\partial y} & \\ & & \dfrac{\partial}{\partial z} \\ & \dfrac{\partial}{\partial z} & \dfrac{\partial}{\partial y} \\ \dfrac{\partial}{\partial z} & & \dfrac{\partial}{\partial x} \\ \dfrac{\partial}{\partial y} & \dfrac{\partial}{\partial x} & \end{bmatrix} \tag{2.4.5}$$

These operators in Eq. 2.4.3 can be presented as:

$$-ik_{i\alpha} = -ikn_{i\alpha} = -ik\begin{bmatrix} n_x & & & & n_z & n_y \\ & n_y & & n_z & & n_x \\ & & n_z & n_y & n_x & \end{bmatrix} \tag{2.4.6}$$

$$-ik_{\beta j} = -ikn_{\beta j} = -ik \begin{bmatrix} n_x & & \\ & n_y & \\ & & n_z \\ & n_z & n_y \\ n_z & & n_x \\ n_y & n_x & \end{bmatrix} \tag{2.4.7}$$

The equation of wave propagation with $\vec{F} = 0$ becomes:

$$k^2 \left(n_{i\alpha} c_{\alpha\beta} n_{\beta j} \right) V_j = k^2 \Gamma_{ij} V_j = \rho \omega^2 V_i \tag{2.4.8}$$

We found again Christoffel's equation Eq. 2.1.5 with Christoffel's matrix:

$$\Gamma_{ij} = n_{i\alpha} c_{\alpha\beta} n_{\beta j} \tag{2.4.9}$$

Let consider the propagation of a P-wave and a S-wave in isotropic solid. Since all directions in such a solid are equivalent we can pose $\vec{n} = n_3 \vec{z}$. Equation 2.4.8 can be presented in the form:

$$k^2 \begin{bmatrix} c_{44} & & \\ & c_{44} & \\ & & c_{11} \end{bmatrix} \begin{bmatrix} V_1 \\ V_2 \\ V_3 \end{bmatrix} = \rho \omega^2 \begin{bmatrix} V_1 \\ V_2 \\ V_3 \end{bmatrix} \tag{2.4.10}$$

From Eq. 2.4.10 we get three independent equations:

$$\begin{aligned} k^2 c_{44} V_1 &= \rho \omega^2 V_1 \\ k^2 c_{44} V_2 &= \rho \omega^2 V_2 \\ k^2 c_{11} V_3 &= \rho \omega^2 V_3 \end{aligned} \tag{2.4.11}$$

An S-wave propagating along z-axis with polarization along x-axis is given by:

$$\vec{V}_{S^x} = \vec{n}_x V e^{i(\omega t - kz)} \tag{2.4.12}$$

The y-axis polarized S-wave will be:

$$\vec{V}_{S^y} = \vec{n}_y V e^{i(\omega t - kz)} \tag{2.4.13}$$

Both Eq. 2.4.12 and Eq. 2.4.13 satisfy the first dispersion relation in Eq. 2.4.11. The general solution for a S-wave propagating in any direction is:

$$\vec{V}_S = \vec{a} V e^{i(\omega t - k\vec{n} \cdot \vec{r})} \text{ with } \vec{a} \cdot \vec{n} = 0 \text{ as well as } \vec{V}_{S'} = \vec{a} \times \vec{n} V e^{i(\omega t - k\vec{n} \cdot \vec{r})}$$

A P-wave propagating along z-axis can be presented in the form:

$$\vec{V}_P = \vec{n}_z V e^{i(\omega t - kz)} \tag{2.4.14}$$

Equation 2.4.14 satisfies the third dispersion relation in Eq. 2.4.11 and the general solution of P-wave propagation is written as:

$$\vec{V}_P = \vec{n} V e^{i(\omega t - k\vec{n}\cdot\vec{r})} \tag{2.4.15}$$

The propagation of both P- and S-waves is lossless. Let see what happens if the medium of propagation is absorbing elastic energy. The amplitude of the elastic wave that propagates in a medium with elastic losses will steadily decrease and its energy will be absorbed or dissipate by the medium. In this case for x-axis polarized wave propagating along y-axis in a medium with attenuation coefficient α the displacement can be written as:

$$\vec{u} = \vec{n}_x e^{-\alpha y} e^{i(\omega t - ky)} \tag{2.4.16}$$

The strain will be:

$$S_{12} = \frac{1}{2}\frac{\partial u_1}{\partial y} = -i\frac{k - i\alpha}{2} e^{i\omega t} e^{-i(k - i\alpha)y}$$

The elastic losses can be introduced using complex stiffness coefficients with η_{ijkl} being the viscosity tensor of rank 4:

$$c_{\alpha\beta} \to c_{\alpha\beta} + i\omega\eta_{\alpha\beta} \tag{2.4.17}$$

The general equation of propagation Eq. 2.4.3 can be used with Eq. 2.4.17 to get the Hooke's law for a medium with elastic losses:

$$T_{ij} = c_{ijkl} S_{kl} + \eta_{ijkl}\frac{\partial S_{kl}}{\partial t} \tag{2.4.18}$$

If the medium is a cubic crystal with the coordinates are aligned with the cube axes the stress is (Auld 1973):

$$T_{12} = -i(k - i\alpha)(c_{44} + i\omega\eta_{44}) e^{i\omega t} e^{-i(k - i\alpha)y} \tag{2.4.19}$$

The equation of motion for x-polarized wave propagating along y-axis is:

$$\frac{\partial T_{i2}}{\partial y} = \rho\frac{\partial^2 u_i}{\partial t^2}$$

The stress component T_{12} becomes:

$$-i(k - i\alpha)T_{12} = -\omega^2 \rho u_1 \tag{2.4.20}$$

With Eq. 2.4.16 and Eq. 2.4.19 into Eq. 2.4.20 we get:

$$(k - i\alpha^2)(c_{44} + i\omega\eta_{44}) = (k^2 - \alpha^2 - i2\alpha k)(c_{44} + i\omega\eta_{44}) = \rho\omega^2 \quad (2.4.21)$$

The real and imaginary parts of Eq. 2.4.20 are:

$$c_{44}(k^2 - \alpha^2) + 2\alpha k\omega\eta_{44} = \rho\omega^2$$

$$i((k^2 - \alpha^2)\omega\eta_{44} - 2\alpha k c_{44}) = 0 \quad (2.4.22)$$

Solving Eq. 2.4.21 for α and eliminating k we get the expression of the attenuation coefficient of the medium of propagation (Auld 1973):

$$\alpha = \left[\frac{\rho\omega^2}{2c_{44}} \left(\frac{1}{\left(1 + \left(\frac{\omega\eta_{44}}{c_{44}}\right)^2\right)^{1/2}} - \frac{1}{1 + \left(\frac{\omega\eta_{44}}{c_{44}}\right)^2} \right) \right]^{1/2} \approx \frac{\omega^2}{2} \frac{\eta_{44}}{c_{44}} \left(\frac{\rho}{c_{44}}\right)^{1/2} \quad \text{with} \quad \left(\frac{\omega\eta_{44}}{c_{44}}\right)^2 \ll 1$$

$$(2.4.23)$$

In the case of an isotropic lossless medium of propagation the stiffness matrix is given by Eq. 1.5.1.3 where $c_{44} = (c_{11} - c_{12})/2$. In the case of an isotropic medium of propagation with viscous losses we have $\eta_{44} = (\eta_{11} - \eta_{12})/2$, respectively. Following the same pattern as in the case of propagation in a lossless medium we get the Christoffel's matrix:

$$k^2(c_{44} + i\omega c_{44}) = \rho\omega^2$$

and

$$k^2(c_{11} + i\omega c_{11}) = \rho\omega^2$$

Equation 2.4.11 and Eq. 2.4.22 give:

$$\alpha_S = \frac{\omega^2}{2} \frac{\eta_{44}}{c_{44}} \left(\frac{\rho}{c_{44}}\right)^{1/2} = \frac{1}{2} \frac{\omega}{Q_{44}^S} \quad \text{where} \quad Q_{44}^S = \frac{c_{44}}{\omega\eta_{44}} \quad (2.4.24)$$

Equation 2.4.24 is the attenuation coefficient of a S-waves propagation in a isotropic medium with losses. For the P-wave we have respectively:

$$\alpha_P = \frac{1}{2} \frac{\omega}{Q_{11}^P} \quad \text{with} \quad Q_{11}^P = \frac{c_{11}}{\omega\eta_{11}} \quad (2.4.25)$$

Q is called elastic Q-factor which is inversely proportional to the attenuation coefficient. The higher the Q-factor the lower elastic losses are. The Q-factor is also inversely proportional to the frequency. That means that the Q-factor is very high a low frequencies. For example for an S-wave propagating in a rock medium with parameters $c_{44} = 10^{10} N/m^2$ and $\eta_{44} = 3 \times 10^{-4} N.sec/m^2$ the Q-factor is $Q = 0.3 \times 10^{14}$ at a frequency of 1 Hz. Since the Earth's lithosphere is formed not only of rocks but also of many other energy absorbing components, the measured average Q-factor is much lower—about 5×10^2 (Sato 1998).

Equation 2.4.16 shows that the amplitude of an elastic wave that propagates in a medium with losses decreases exponentially with the travel distance. Between two points y_1 and y_2 the amplitude will decrease by a factor of $e^{\alpha(y_2-y_1)}$. Since $\alpha(y_2 - y_1)$ is dimensionless the attenuation coefficient is measured per unit length, i.e., in nepers/meter (Np/m) or in decibels/meter (dB/m).

$$1 Np/m = \frac{20}{\log_e 10} dB \approx 8.686 dB$$

$$1 dB/m = \frac{1}{\log_{10} e} Np \approx 0.1151 Np \tag{2.4.26}$$

The attenuation coefficient is proportional to ω^2. The Q-factors of a quartz resonators operating in the MHz-range vary between 10^4 and 10^6. The frequencies of seismic waves are about 6 orders of magnitude lower. The Q-factor values measured in the lithosphere are about 500 for S-waves and 1100 for P-waves at 1 Hz (Sato and Fehler 1998). That means P- and S-waves can still travel thousands of miles before vanishing. As it will be discussed later the main factor of seismic wave attenuation in the lithosphere is not elastic energy absorption but wave scattering.

Seismic wave frequency usually is in the range 0.5 Hz–50 Hz. Why seismic waves are generated in this frequency range? Why they are not propagating at higher frequencies, for example, 1 MHz or at lower frequency like 0.01 Hz? Why do they look like waves that have been going through a bandpass filter? When discussing the Fourier theorem we noticed an important corollary stating that short signals (pulses) in the time domain occupy a wide bandwidth and wide signals (pulses) in the time domain occupy a narrow bandwidth. A seismic chock in the focus of the earthquake causes usually a short pressure pulse. Therefore all seismic waves that originate from this pulse should have frequencies in a wide frequency range. Indeed, they do get generated in a much wider spectrum than 0.5 Hz–50 Hz, but they are not 'relevant' to the earthquake's effects because they get bandpass filtered. At 1 MHz, for example, the wavelength would be just $4.10^{-3} m$ for a velocity of propagation of 4,000 m/sec. At such wavelength the wave will get scattered from random small domains in the heterogeneous medium of the Earth's crust or mantra according to Rayleigh scattering theory. As a result of the intensive scattering the wave will get evanescent at a very short distance and become part of the Earth's vibration noise. For a very long wavelength of 0.01 Hz the wavelength is 400,000 m for a velocity of 4,000 m/sec. The amplitude of a seismic wave with such long wavelength will be barely detected by modern seismometers. The 151B Rev Tec seismometer, for example, with three independent sensors, is capable to detect seismic signals in the range of 0.0083 Hz to 50 Hz.

For a bulk wave propagating with a velocity of 4,000 m/sec we have calculated that in the range 0.5 Hz–50 Hz the wavelength of a seismic wave is between 80 m and 8,000 m and for a surface wave propagating with a velocity of 2,000 m/sec the wavelength is between 40 m and 4,000 m. In the most destructive frequency range of 1 Hz–10 Hz the wavelength for body waves is in the range between 4,000 m–400 m and for surface waves between 2,000 m–200 m. The size of the rock slabs forming the heterogeneous structure of the Earth's crust where most of the seismic activities take place act as low-pass mechanical filter to seismic waves that letting pass only waves with frequencies lower than the cutoff frequency and attenuates waves with frequencies higher than the cutoff frequency. Only seismic waves in that frequency range can travel at long distances causing earthquake's effect. All other waves become part of the Earth's acoustic noise (or coda)—continuous small vibrations that have no impact on manmade constructions unless they get amplitude modulated or amplified by nonlinear phenomena. An analogy of this in optics is the blue color of the sky and its red color around the sun at sunset. The blue color of the sky is caused by much stronger Rayleigh scattering of blue light than other visible light from airborne gas molecules which size is much smaller than the wavelength. The blue light has the shortest wavelength of the visible spectrum. Red, yellow light and other components of the visible spectrum get less scattered by the gas molecules than the blue light because of their longer wavelength. They just pass unperturbed through the blue air. This is also the reason of the yellow/red color around the sun at sunset. As sunlight passes through the atmosphere, its blue component is Rayleigh scattered strongly by atmospheric gases but the longer wavelength (e.g., red/yellow) components are not. The sunlight arriving directly from the sun therefore appears to be slightly yellow while the light scattered through rest of the sky appears blue. During sunrises and sunsets, the Rayleigh scattering effect is much more noticeable due to the larger volume of air through which sunlight passes.

Despite that elastic wave scattering from random objects in heterogeneous media is more complex phenomenon than scattering of visible light the analogy is useful. Exact solution to Rayleigh scattering problem for spherical acoustic waves scattered from spherical obstacle has been reported (Godin 2013). Despite the complexity of the general problem of elastic wave scattering from random objects it could be claimed that short-wave seismic waves get scattered from a great number of randomly dispersed small domains in the Earth's crust and mantle with size comparable to the wavelength, whereas long-wave seismic waves pass long distances undisturbed.

2.5 Nonlinear elastic wave propagation

The dependence of velocity of propagation on pressure indicates that wave propagation in rocks is highly nonlinear. When Landau moduli *A*, *B*, and *C* are measured in rocks they are found to be several orders of magnitude higher than the moduli of Al, water and other constituents of the lithosphere. These observations are at odds with the linear elasticity theory as well as with the basic starting points of the theory of nonlinear elasticity. A new theory has been proposed relating stress-strain measurements to density ρ and use ρ to find the behaviors of nonlinear elastic waves in rocks (McCall 1996).

However, there are also other factors that contribute to the complexity of the problem. Rocks contain pockets of encapsulated elastic energy. During the propagation of seismic waves these pockets of higher elastic energy density act as additional seismic energy sources.

On September 19, 1985 Mexico City was struck by an 8.1-magnitude earthquake that caused the deaths of more than 10,000 people and devastated the whole area. The earthquake occurred in the Pacific Ocean, off the coast of the Mexican state of Michoacán, a distance of more than 350 km from the city, in the Cocos Platesubduction zone, specifically in a section of the fault line known as the Michoacán seismic gap. Why seismic waves coming from so far were so devastating? Two reasons were given to explain it. One was the unexpected way the Mexico City's underlying layers of soft clay soil behaved during the tremor. This weak soil transmitted much more ground movement than engineers and planners had ever expected for such a distant earthquake (Peterson 1986). The second reason was the resonance in the lakebed sediments and the long duration of the shaking (Murillo 1995). The most damaged buildings were from 6 to 15 stories in height. These buildings tended to resonate most in the frequency band of the lakebed motions. Many buildings had their upper floors collapsed while the lower floors were left relatively undamaged. Explained in terms of nonlinear elastic wave propagation when seismic waves propagate through soft superficial alluvial layers or scatter on strong topographic irregularities, refraction or scattering phenomena may strongly increase the amplitude of the ground motion. At the scale of an alluvial basin, seismic effects involve various phenomena, such as wave trapping, resonance of the whole basin, propagation in heterogeneous media, and the generation of surface waves at the basin edges (Brûlé 2014).

Strong nonlinear phenomena of finite-amplitude elastic waves propagating in solids with low attenuation causing waveform distortion and growth of higher harmonics have been reported in the middle of the 20th century (Cedroits and Krasil'nikov 1963).

The cause of higher harmonics growth (Fig. 2.5.1a) and appearance of combination frequency waves (Fig. 2.5.1b) is the nonlinearity of equation of state

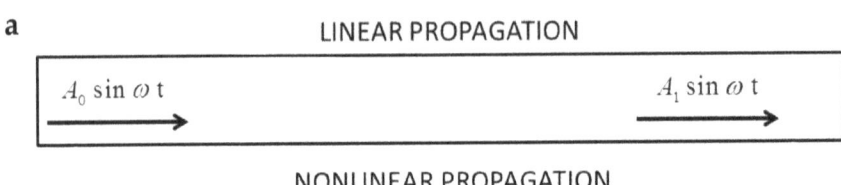

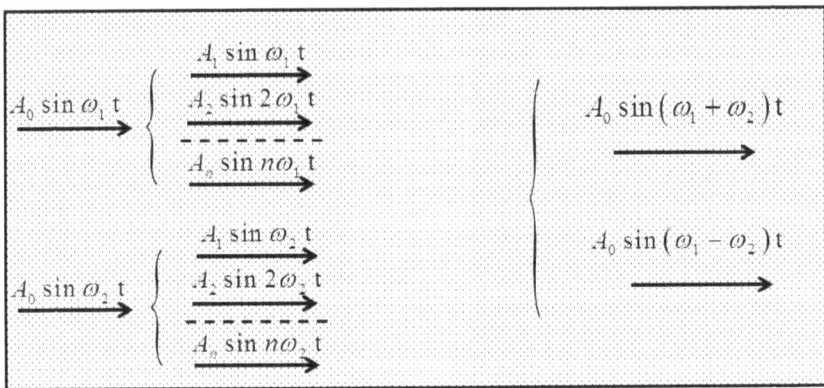

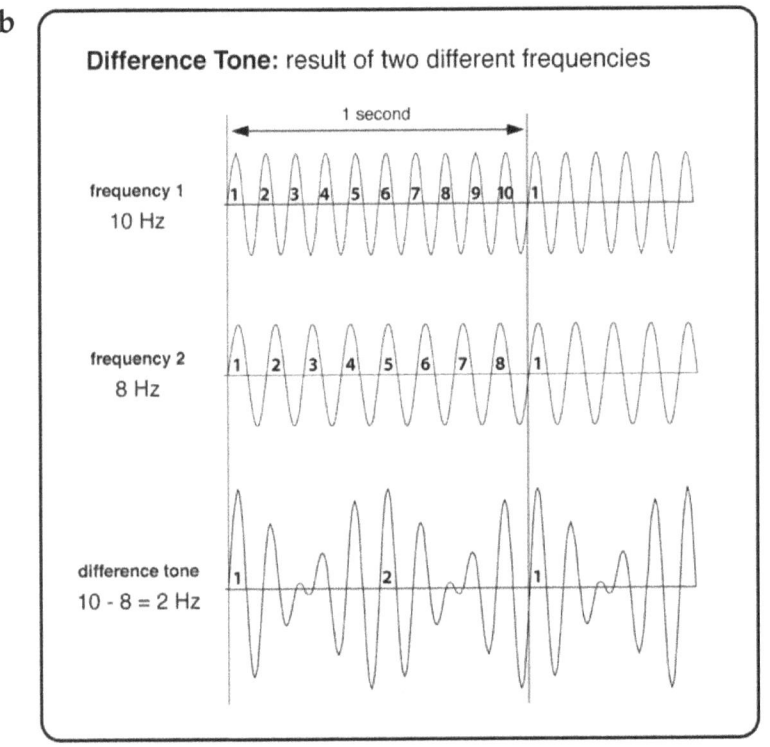

Fig. 2.5.1. (a) Nonlinear generation of harmonics; (b) Example of combination frequency waves generation.

of the medium of propagation and the geometrical characteristics of the finite deformations. In acoustics this phenomenon is called beat frequency sound. When two tones are close in frequency, the difference in frequency generates a beating. Interference between the two tones occurs resulting in periodical variation of the volume. In Chapter 1 we have discussed this phenomenon and the amplitude modulation that it produces. Musical instruments tuning that can produce sustained tones, beats can readily be recognized. A trembling effect (tremolo) is produced because the volume varies as the sounds alternately interfere constructively and destructively. As the two tones gradually approach their frequencies, the beating slows down and may become imperceptible. This effect can be seen in Fig. 2.5.1b. It can be seen at any seismogram as well (for example Fig. 1.1.1.2). We can call it 'seismic beat'.

The nonlinearity of the equation of state of the medium of propagation leads to the appearance of terms containing squares and cubes of the displacements u_{ij} in the expression of elastic energy density. The nonlinearity due to the finite-amplitude deformations leads to nonlinear relationship between the strain tensor components and the derivatives of the displacement components.

We have noted already that materials that are subjected to large strain behave in a nonlinear way. This is due either to nonlinearity of the material itself or to strain which can be for example due to a propagating high-amplitude elastic wave. The wave front of a small-amplitude elastic wave in a heterogeneous medium of propagation gets quickly deformed and the wave loses its energy similarly to as it were propagating in a high-attenuation dissipative medium. However, elastic heterogeneous media can support steadily high-amplitude elastic waves due to the strong nonlinearity of the material and to the wave dispersion caused by scattering with the heterogeneous structure. This is exactly what happens in the rock.

2.5.1 Nonlinear corrections to Hooke's law

Let' consider an isotropic solid with no elastic energy dissipation and dispersion. A solid is considered to be elastic if after being deformed or strained by external forces it returns to its initial equilibrium shape when the forces stop acting due to internal restoring forces. Using Fig. 1.5.1.3 we have found that the displacement $d\vec{u}$ can be expressed using the displacement gradient tensor $\partial \vec{u}/\partial x_i$ resulting from strain as:

$$d\vec{u} = \frac{\partial \vec{u}}{\partial x_i} dx_i$$

(2.5.1.1)

Equation 2.5.1.1 is the second term in the Taylor's series of the displacement u_i:

$$u_i(x) = u_i(x=0) + \left(\frac{\partial u_i}{\partial x_j}\right)_{x=0} dx_j + \frac{1}{2}\left(\frac{\partial^2 u_i}{\partial x_j \partial x_k}\right)_{x=0} dx_i dx_j + \ldots =$$

$$u_i(x=0) + \frac{1}{2}\left(\frac{\partial u_i}{\partial x_j} + \frac{\partial u_j}{\partial x_i}\right) dx_i + \frac{1}{2}\left(\frac{\partial u_i}{\partial x_j} - \frac{\partial u_j}{\partial x_i}\right) dx_j + \ldots$$

(2.5.1.2)

The symmetric and antisymmetric parts of the displacement gradient tensor are:

$$\varepsilon_{ij} = \frac{1}{2}\left(\frac{\partial u_i}{\partial x_j} + \frac{\partial u_j}{\partial x_i}\right) \text{ and } \sigma_{ij} = \frac{1}{2}\left(\frac{\partial u_i}{\partial x_j} - \frac{\partial u_j}{\partial x_i}\right)$$

From Eq. 1.5.1.9 we have:

$$T_{ij} = c_{ijkl}\frac{\partial u_i}{\partial x_k}$$

Developed in Taylor's series the stress as a function of the strain is given by:

$$T_{ij}(S_{kl}) = T_{ij}(S_{kl}=0) + \left(\frac{\partial T_{ij}}{\partial S_{kl}}\right)_{S_{kl}=0} S_{kl} + \frac{1}{2}\left(\frac{\partial^2 T_{ij}}{\partial S_{kl}\partial S_{mn}}\right)_{S_{kl}=0,S_{mn}=0} S_{kl}S_{mn} + \cdots \text{ (2.5.1.3)}$$

Since $T_{ij}(0) = 0$ we get Hooke's law (Eq. 1.5.1.1) with the second-order stiffness constant given by the following expression:

$$c_{ijkl} = \left(\frac{\partial T_{ij}}{\partial S_{kl}}\right)_{S_{kl}=0}$$

(2.5.1.4)

The third-order stiffness constants are respectively:

$$c_{ijklmn} = \frac{1}{2}\left(\frac{\partial^2 T_{ij}}{\partial S_{kl}\partial S_{mn}}\right)_{S_{kl}=0;S_{mn}=0}$$

(2.5.1.5)

The nonlinear Hooke's law can be written in the form:

$$T_{ij} = c_{ijkl}S_{kl} + c_{ijklmn}S_{kl}S_{mn} + \ldots$$

(2.5.1.6)

The elastic strain energy density is:

$$dE_p = T_{ij}dS_{ij}$$

(2.5.1.7)

From Eq. 2.5.1.7 we get the stress in the form:

$$T_{ij} = \frac{\partial E_P}{\partial \left(\partial u_i / \partial x_j \right)} \tag{2.5.1.8}$$

Developed to the third order in strain the elastic energy is given by the following expression (Brugger 1964):

$$E_P = \frac{1}{2} c_{ijkl} S_{ij} S_{kl} + \frac{1}{6} c_{ijklmn} S_{ij} S_{kl} S_{mn} + \dots \tag{2.5.1.9}$$

2.5.2 Five-constant theory

We will still be considering an isotropic solid with no elastic energy dissipation and dispersion. If the total energy (kinetic + potential) in unit volume is $E = E_k + E_p$ the Lagrangian is defined by: $L = E_k - E_p$. The mass at the end of the spring shown in Fig. 1.5.1.1 that we used to discuss the Hooke's law has a kinetic energy that is equal to $E_k = m\dot{x}^2/2$ and a potential energy is equal to $E_p = kx^2/2$. Therefore, Lagrangian can be presented as:

$$L = \frac{1}{2} m\dot{x}^2 - \frac{1}{2} kx^2 \tag{2.5.2.1}$$

The equation of motion of the mass at the end of the spring comes from Newton's law $F = ma$:

$$m\ddot{x} = -kx \tag{2.5.2.2}$$

With the Lagrangian the equation of motion of the mass attached at the end of the spring can be written as:

$$\frac{d}{dt}\left(\frac{\partial L}{\partial \dot{x}} \right) = \frac{\partial L}{\partial x} \tag{2.5.2.3}$$

This equation is called Euler-Lagrange equation. Since the equation of motion and Euler-Lagrange equation are describing the same phenomenon, the Euler-Lagrange equation should be valid in all coordinate systems. If, instead of mass at the end of the spring, we consider a more general case of a mass moving in a field with potential energy $E_p(x)$ the Lagrangian will be:

$$L = \frac{1}{2} m\dot{x}^2 - E_p(x) \tag{2.5.2.4}$$

The Euler-Lagrange equation becomes:

$$m\ddot{x} = -\frac{dE_p}{dx} \tag{2.5.2.5}$$

The force applied to the mass is therefore equal to:

$$F = -\frac{dE_p}{dx}$$ (2.5.2.6)

In the more general case if a mass is moving in a potential $V(x)$ the Euler-Lagrange equation is:

$$m\ddot{x} = -\frac{dV(x)}{dx}$$ (2.5.2.7)

Therefore, dV/dx is the force applied to the mass moving in the potential $V(x)$. We find again Newton's law $F=ma$ in one-dimensional systems of coordinates. In three-dimensional Cartesian coordinates we have accordingly:

$$L = \frac{1}{2}m\left(\dot{x}^2 + \dot{y}^2 + \dot{z}^2\right) - V\left(x, y, z\right)$$ (2.5.2.8)

This equation can be presented also in the form:

$$m\ddot{\vec{r}} = -\nabla E_p;\ \ \begin{array}{c}\vec{r} = \vec{x} + \vec{y} + \vec{z}\\ F = -\nabla E_p\end{array} \Rightarrow F = -\nabla E_p \Leftrightarrow \vec{F} = m\vec{a}$$ (2.5.2.9)

The above once again conforms that Euler-Lagrange equation and Newton's law are describing the same phenomenon. The Euler-Lagrange equation can be expressed in any other coordinates q_i in the same way it is expressed in Cartesian ones if $q_i = q_i(x_1, x_2, ...x_N; t)$ and reversibly $x_i = x_i(q_1, q_2, ...q_N; t)$, i.e.:

$$\frac{d}{dt}\left(\frac{\partial L}{\partial \dot{q}_n^2}\right) = \frac{\partial L}{\partial q_n};\ 1 \leq n \leq N$$ (2.5.2.10)

We will consider now the propagation of a high-amplitude elastic wave in an isotropic solid. The equation of propagation of a linear elastic wave (Eq. 2.1.1) in isotropic solid with $\frac{\partial^2 u_i}{\partial t^2} = \ddot{u}_i$ is:

$$\rho\ddot{u}_i = \frac{\partial T_{ij}}{\partial x_j}$$ (2.5.2.11)

From 2.5.1.8 in adiabatic approximation we have:

$$T_{ij} = \left(\frac{\partial E_p}{\partial u_{ij}}\right)$$ (2.5.2.12)

The relation between the displacement and the strain tensor T_{ij} given by:

$$T_{ij} = \frac{1}{2}\left(\frac{\partial u_i}{\partial x_j} + \frac{\partial u_j}{\partial x_i} + \frac{\partial u_k}{\partial x_i}\frac{\partial u_k}{\partial x_j}\right) \qquad (2.5.2.13)$$

On the other hand we have found in Chapter 2 that in an isotropic solid we have:

$$T_{ij} = c_{ijkl}S_{kl} = \left(\lambda\delta_{kl} + 2\mu\delta_{ik}\delta_{il}\right)S_{kl} \qquad (2.5.2.14)$$

Since

$$T_{ii} = \lambda\left(S_{11} + S_{22} + S_{33}\right) + 2\mu S_{ii} \Rightarrow T_{ij} = c_{ijkl}S_{kl} = \mu\left(\delta_{ik}\delta_{jl} + \delta_{il}\delta_{jk}\right)S_{kl} = 2\mu S_{ij}; \left(i \neq j\right)$$

$$(2.5.2.15)$$

T_{ii} and T_{ij} can be put together in the same expression if $i \neq j$ which corresponds to a simple volume dilatation uniformly in all directions:

$$T_{ij} = \lambda S_{ii} + 2\mu S_{ij} = \kappa S_{ii} + 2\mu\left(S_{ij} - \frac{1}{3}S_{ii}\right); \text{ with } \kappa = \lambda + \frac{2}{3}\mu \quad (2.5.2.16)$$

The equation of propagation can be rewritten as:

$$\rho\ddot{u}_i = \left(\lambda + \mu\right)\frac{\partial^2 u_j}{\partial x_i \partial x_j} + \mu\frac{\partial^2 u_i}{\partial x_j^2} \qquad (2.5.2.17)$$

Equation 2.5.1.17 leads to the solutions of a longitudinal and a shear elastic waves propagating with velocities given by:

$$V_P = \sqrt{\left(\lambda + 2\mu\right)/\rho} \text{ and } V_S = \sqrt{\mu/\rho} \qquad (2.5.2.18)$$

Equation 2.5.2.17 is still the small-amplitude linear elastic wave equation of propagation. This means that the amplitude has been decomposed in a Taylor's series and all high-order terms have been neglected. In the case of a high amplitude elastic wave we have to take into account these high-order terms as well.

If $L(q, \dot{q}, t)$ is the Lagrangian describing a mechanical system we can define the integral $A = \int L(q, \dot{q}, t)dt$ called action integral and with $A=0$ we have the Hamilton's principle of stationary action in mechanics corresponding to the Fermat principle in optics.

The Euler-Lagrange equation is presented by:

$$\frac{d}{dt}\left(\frac{\partial L}{\partial \dot{q}_n^2}\right) - \frac{\partial L}{\partial q_n} = 0 \qquad (2.5.2.19)$$

To any velocity there is a corresponding momentum p_i:

$$p_i = \frac{\partial L}{\partial \dot{q}_i} \qquad (2.5.2.20)$$

We can define the Hamiltonian operator in a similar way as the Lagrangian operator:

$L(q_i,\dot{q}_i,t) \rightarrow H(p_i,\dot{p}_i,t)$ where $H(p_i,\dot{p}_i,t) = \sum_j \dot{q}_j p_j - L(q_i,\dot{q}_i,t)$ using Legendre transform.

From the Euler-Lagrange Eq. 2.5.2.19 we get $H = E_k + E_p$ the Hamilton's equations:

$$\frac{\partial H}{\partial q_i} = -\dot{p}_i \text{ and } \frac{\partial H}{\partial p_i} = \dot{q}_i \qquad (2.5.2.21)$$

If the action integral depends on q_i and t, i.e.,

$$A(q_i,t) = \int L dt = \int (p_i \dot{q}_i - H) dt = \int dA(q_i,t) dt \rightarrow p_i = \frac{\partial A}{\partial q_i} \quad (2.5.2.22)$$

Equation 2.5.2.22 leads to the Hamilton-Jacobi equation:

$$\frac{\partial A}{\partial t} + H\left(q_i, \frac{\partial A}{\partial q_i}, t\right) = 0 \qquad (2.5.2.23)$$

In third order nonlinear approximation the elastic potential energy density (=the elastic energy per unit volume) of an isotropic sold is given by (Landau and Lifshitz 1959):

$$E_P = \mu u_{ij}^2 + \left(\frac{K}{2} - \frac{\mu}{3}\right)u_{ll}^2 + \frac{1}{3}Au_{ij}u_{il}u_{jl} + Bu_{ij}^2 u_{ll} + \frac{1}{3}Cu_{ll}^2 \qquad (2.5.2.24)$$

Replacing Eq. 1.5.1.6 in Eq. 2.5.2.24 we obtain for an isotropic solid:

$$E_P = \frac{\mu}{4}\left(\frac{\partial u_i}{\partial x_j} + \frac{\partial u_j}{\partial x_i}\right)^2 + \left(\frac{K}{2} - \frac{\mu}{3}\right)\left(\frac{\partial u_l}{\partial x_l}\right)^2 + \left(\mu + \frac{A}{4}\right)\frac{\partial u_i}{\partial x_j}\frac{\partial u_i}{\partial x_i}\frac{\partial u_l}{\partial x_j} + \qquad (2.5.2.25)$$

$$\left(\frac{B}{2} + \frac{K}{2} - \frac{\mu}{3}\right)\frac{\partial u_l}{\partial x_i}\frac{\partial u_l}{\partial x_i}\left(\frac{\partial u_i}{\partial x_j}\right)^2 + \frac{A}{12}\frac{\partial u_i}{\partial x_j}\frac{\partial u_i}{\partial x_i}\frac{\partial u_l}{\partial x_j} + \frac{B}{2}\frac{\partial u_i}{\partial x_j}\frac{\partial u_j}{\partial x_i}\frac{\partial u_l}{\partial x_l} + \frac{C}{3}\left(\frac{\partial u_l}{\partial x_l}\right)^3$$

Equation 2.5.1.9 is the general expression for the potential energy in an anisotropic solid, so Eq. 2.5.2.25 is derived from it. Third-order moduli A, B, C (Landau moduli) are included in Eq. 2.5.2.25 together with the moduli of linear uniform volume compression as well as shear moduli. The five moduli A, B, C, K, and μ in Eq. 2.5.2.25 characterize the nonlinear deformation of the isotropic solid known as 'five-constant theory'.

The linear equation of motion (Eq. 2.1.1) can be extended to a nonlinear case by inserting higher-order displacement terms and using stress tensor that includes quadratic terms. In Chapter 1 we defined the stress tensor in Cartesian coordinates called 'true stress' or Cauchy stress. In Lagrangian coordinates Cauchy tensor cannot be directly defined. By introducing a fictive action force related to the actual force in a systematic way we are able to define a new appropriate measure of stress in the new reference coordinates called Piola-Kirchhoff that can be used to go back and calculate the true Cauchy stress. If N_{ij} is a Piola-Kirchhoff tensor the equation of motions Eq. 2.1.1 can be presented using in the form:

$$\rho \ddot{u}_i = \frac{\partial N_{ij}}{\partial x_j} \tag{2.5.2.26}$$

The tensor N_{ij} is defined by:

$$N_{ij} = \frac{\partial E_p}{\partial (\partial u_i / \partial x_j)} \tag{2.5.2.27}$$

Using Eq. 2.5.2.26 and 2.5.2.27 the following expression for the tensor N_{ij} can be obtained (Thurston 1966):

$$N_{ij} = \left(\mu + \frac{A}{4} \right) \left(\frac{\partial u_l}{\partial x_i} \frac{\partial u_l}{\partial x_j} + \frac{\partial u_l}{\partial x_i} \frac{\partial u_j}{\partial x_l} + \frac{\partial u_l}{\partial x_j} \frac{\partial u_i}{\partial x_l} \right) + \left(B + K - \frac{2\mu}{3} \right) \frac{\partial u_l}{\partial x_j} \frac{\partial u_i}{\partial x_l} + $$
$$B \frac{\partial u_j}{\partial x_i} \frac{\partial u_l}{\partial x_l} + \frac{A}{4} \frac{\partial u_j}{\partial x_l} \frac{\partial u_l}{\partial x_i} + \left[\left(\frac{B}{2} + \frac{K}{2} - \frac{\mu}{3} \right) \frac{\partial u_l}{\partial x_k} \frac{\partial u_l}{\partial x_k} + \frac{B}{2} \frac{\partial u_l}{\partial x_k} \frac{\partial u_k}{\partial x_l} + C \left(\frac{\partial u_l}{\partial x_l} \right)^2 \right] \delta_{ij}$$

With Eq. 2.5.2.25 and Eq. 2.5.2.27 we get the equation of motion:

$$\rho \frac{\partial^2 u_i}{\partial t^2} - \mu \frac{\partial^2 u_i}{\partial x_l^2} - \left(K + \frac{\mu}{3} \right) \frac{\partial^2 u_k}{\partial x_k \partial x_i} = F_i \tag{2.5.2.28}$$

In Eq. 2.5.2.28 F_i is presented by:

$$F_i = \left(\mu + \frac{A}{4} \right) \left(\frac{\partial^2 u_k}{\partial x_j^2} \frac{\partial u_k}{\partial x_i} + \frac{\partial^2 u_k}{\partial x_j^2} \frac{\partial u_i}{\partial x_k} + 2 \frac{\partial^2 u_i}{\partial x_k \partial x_j} \frac{\partial u_k}{\partial x_j} \right) + \left(K + \frac{\mu}{3} + \frac{A}{4} + B \right) \left(\frac{\partial^2 u_j}{\partial x_i \partial x_j} \frac{\partial u_k}{\partial x_k} + \frac{\partial^2 u_j}{\partial x_j \partial x_k} \frac{\partial u_i}{\partial x_k} \right) - $$
$$\left(K + \frac{2\mu}{3} + B \right) \frac{\partial^2 u_i}{\partial x_j^2} \frac{\partial u_k}{\partial x_k} + \left(\frac{A}{4} + B \right) \left(\frac{\partial^2 u_j}{\partial x_i \partial x_j} \frac{\partial u_k}{\partial x_i} + \frac{\partial^2 u_k}{\partial x_i \partial x_j} \frac{\partial u_k}{\partial x_k} \right) + (B + 2C) \frac{\partial^2 u_j}{\partial x_i \partial x_j} \frac{\partial u_k}{\partial x_k} \tag{2.5.2.29}$$

For a longitudinal elastic wave Eq. 2.5.2.28 becomes:

$$\frac{\partial^2 u_L}{\partial x^2} - \frac{1}{V_L^2}\frac{\partial^2 u_L}{\partial t^2} = -2\xi_L \frac{\partial^2 u_L}{\partial x^2}\frac{\partial u_L}{\partial x} \qquad (2.5.2.30)$$

where $\xi_L = \frac{3}{2} + (A + 3B + C)\frac{1}{\rho V_L^2}$.

In the case of a shear wave no second-order term appears, so the linear version of the equation of motion still can be used, however, the third-order term of the displacement is present which complicates the calculation.

For example for a longitudinal wave propagating along the x_1-axis the only nonzero displacement is along the x_1-axis. The only nonzero component of the stress tensor is $N_{11} \neq 0$. Accordingly with Eq. 2.5.1.8 the stress tensor is given by:

$$N_{xx} = \frac{\partial E_P}{\partial (\partial u_1 / \partial x_1)} = \beta \frac{\partial u_x}{\partial x_1} + \chi \left(\frac{\partial u_x}{\partial x_1}\right)^2 \qquad (2.5.2.31)$$

In Eq. 2.5.2.31 we have:

$$\beta = K + \frac{4}{3}\mu,\ \chi = \gamma + 3\beta,\ \text{and}\ \gamma = A + 3B + C \qquad (2.5.2.32)$$

μ is the shear modulus and K is the body longitudinal compression modulus. The nonlinear coefficients χ consists of two components—a coefficient γ to the physical nonlinearity (nonlinearity of the equation of state of the solid) and a coefficient 3β that accounts for the geometrical nonlinearity. The equation of motion including quadratic terms of the P-wave propagation in a medium without dissipation is presented by (Gol'dberg 1960a,b, 1961):

$$\rho_0 \frac{\partial^2 u_1}{\partial t^2} - \beta \frac{\partial^2 u_1}{\partial x_1^2} = \sigma \frac{\partial^2 u_1}{\partial x_1^2}\frac{\partial u_1}{\partial x_1},\ \text{where}\ \sigma = 3\beta + 2\gamma \qquad (2.5.2.33)$$

The solution of the equation of motion Eq. 2.5.2.33 shows that the waveform of the initially sinusoidal wave became distorted during the propagation. In the ratio:

$$\frac{\gamma}{\beta} = \frac{A + 3B + C}{K + \frac{4}{3}\mu} \qquad (2.5.2.34)$$

the parameter β is known and, therefore the parameter γ can be found by measuring the second harmonics amplitude. ρ_0 is the unstrained density.

The amplitude of the second harmonics can be calculated from the solution of Eq. 2.5.2.33:

$$\alpha_2 = -\frac{\left(\dfrac{3}{2}+\dfrac{\gamma}{\beta}\right)\omega\alpha_{01}^2}{2\rho_0 V_P^3}x \qquad (2.5.2.35)$$

In Eq. 2.5.2.35 α_{01} is the amplitude of the P-wave at the source, x is the distance from the source, and V_P is the P-wave velocity of propagation.

From Eqs. 2.5.2.34 and 2.5.2.35 the ratio γ/β is:

$$\frac{\gamma}{\beta} = -\left(\frac{2\rho V_P^2}{\omega\alpha_{01}x}\frac{\alpha_2}{\alpha_{01}}+\frac{3}{2}\right) \qquad (2.5.2.36)$$

The experimental measurements of the second harmonics amplitude of a P-wave provide the value of the nonlinear coefficient γ of an isotropic solid which is a linear combination of Landau moduli A, B, and C. In the case of a solid of cubic symmetry there are six coefficients in the nonlinear Hooke's law for quadratic strain.

2.5.3 Nonlinear propagation in dissipative media

Nonlinear elastic wave propagation involves interactions among waves resulting in growing number of harmonics as well as waves of combined frequencies. In a nondissipative medium the number of harmonics grows fast taking away elastic energy from the fundamental mode. Higher harmonics interact with each other and with the fundamental mode often creating resonance conditions. Harmonics interactions distort the waveform as the wave propagates.

In a dissipative medium of propagation the elastic wave attenuation strongly affects the growth of higher harmonics. If the attenuation is sufficiently high the amplitudes of the higher harmonics are very low and the time waveform of the fundamental mode is not distorted as it propagates. The effect of variations of the relationship between attenuation and frequency affects strongly the higher harmonics growth. As attenuation increases with frequency the growth of harmonics is suppressed so is the energy depletion of fundamental mode (Haran and Cook 1983).

If there is no dispersion and the attention is low Burgers equation is the most suitable to analyze the waveform distortions. Burgers equation is given by:

$$\frac{\partial u}{\partial x} = \vartheta'\frac{\omega_0}{V_0}u\frac{\partial u}{\partial \tau}+\zeta\frac{\partial^2 u}{\partial \tau^2} \qquad (2.5.3.1)$$

In the Burgers equation Eq. 2.5.3.1 $\vartheta' = 1 + B/2A$ with $B/2A$ being the first nonlinear term, ω_0 and V_0 are the angular frequency and velocity of propagation of the fundamental mode, $\tau = \omega_0 t - kx$, and $\zeta = \beta u_0\,\omega_0 / V_0^2 \alpha$ with α being the attenuation coefficient.

Consider an elastic wave in the form:

$$u = \sum_{n=-\infty}^{n=\infty} u_n e^{in(\tau+\phi)} \tag{2.5.3.2}$$

where ϕ is the phase and u_n is the amplitude of the n-th harmonic. We also have $u_n = u_{-n}^*$. Eqs. 2.5.3.2 and 2.5.3.1 give:

$$\frac{\partial u_n}{\partial x} = i\frac{\beta\omega_0}{V_0^2} \sum_{n=-\infty}^{n=\infty} \left[(n-m)u_{n-m}u_m - \alpha_0 n^2 u_n \right] \tag{2.5.3.3}$$

From the linear theory of elasticity we have:

$$u_n = u_{0n} e^{-\alpha_n(x)} \rightarrow \frac{\partial u_n}{\partial x} = -\alpha_n u_n$$

α_n is the attenuation coefficient of the n-th harmonic related to the corresponding frequency. For most nondispersive viscous fluid media $u_n = \alpha_0 n^2$, so a more general expression for isotropic lossy solid where the dependence cannot be expected to be quadratic a good approximation would be:

$$u_n = \alpha_0 n^b \tag{2.5.3.4}$$

Equation 2.5.4.3 can be presented as:

$$\frac{\partial u_n}{\partial x} = i\frac{\beta\omega_0}{V_0^2} \sum_{n=-\infty}^{n=\infty} \left[(n-m)u_{n-m}u_m - \alpha_0 n^b u_n \right] \tag{2.5.3.5}$$

Numerical solutions of Eq. 2.5.3.5 provide the distortion of the waveform as a function of distance at various time moments presented in Fig. 2.5.3.1.

During the nonlinear propagation elastic waves interact with each other giving rise to higher harmonics and waves with combined frequencies. For example two elastic waves with frequencies ω_1 and ω_2 can create a third wave with frequency $\omega_1 + \omega_2 = \omega_3$ which implies $\vec{k}_1 + \vec{k}_2 = \vec{k}_3$ because of the absence of dispersion. In the absence of dispersion we have $k_n = \omega_n / V_n$ and this triplet combination is possible only for collinear waves with parallel wave vectors. However, harmonics with frequencies $n\omega$ can grow with the restriction of collinearity and even cascades of harmonics can be produced (Naugolnykh and Ostrovsky 1998). Energy transfer from the initial fundamental elastic mode toward higher harmonics leads to significant damping of the fundamental wave. In physical acoustics this phenomenon is not desirable and research has been done to avoid it by introducing dispersion or selective damping of harmonics. However, in the case of seismic waves such phenomenon is highly desirable, because the initial fundamental seismic wave would lose its energy

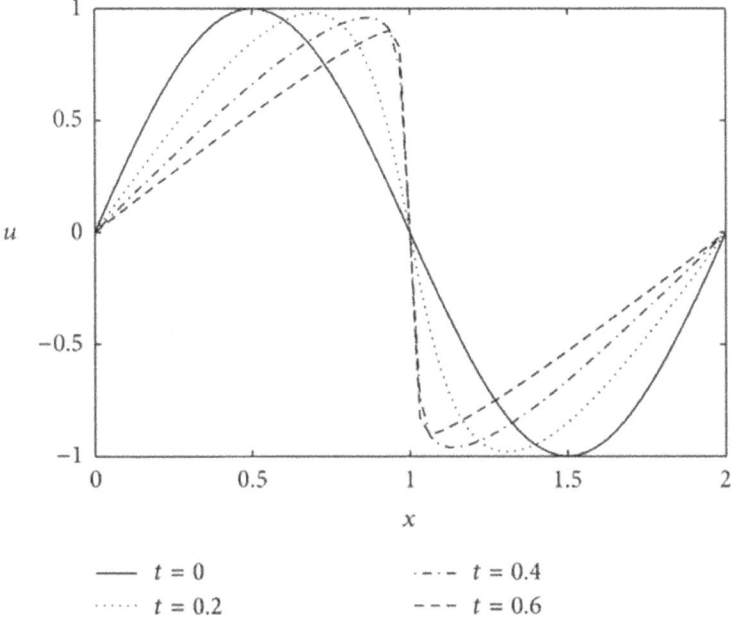

—— $t = 0$	·–·– $t = 0.4$
·········· $t = 0.2$	–––– $t = 0.6$

Fig. 2.5.3.1. Burgers equation solutions at different time moments.

by passing it to higher harmonics. Usually seismic waves are highly dispersive elastic waves and the probability such higher harmonics to grow is very low. Nevertheless, it is worth it to take a look at the phenomena taking place in low-loss nondispersive media.

2.5.4 Propagation in dissipative and dispersive media

As a finite-amplitude wave propagates, its waveform gets distorted due to differences of propagation velocities of its various points. In the presence of dispersion and attenuation these phenomena are even more pronounced. In dispersive medium the formation of a triplet with $\omega_1 + \omega_2 = \omega_3$ does not imply $\vec{k}_1 + \vec{k}_2 = \vec{k}_3$ and these resonance conditions can be satisfied for selected triplets only. However, when satisfied, the interaction will be much stronger.

Dispersion effects are due to various causes: 1) attenuation causing losses of elastic wave energy, 2) media with specific internal spatial or temporal scales, and 3) geometrical dispersion existing in waveguides, bounded systems, and resonators. Selective suppression of spectral components is not excluded.

Assuming that only finite number of spectral components can interact with each other and at certain frequency ranges selective damping occurs. Here it is appropriate to use the nonlinear equation (Rudenko 1983):

$$\frac{\partial u}{\partial x} = \xi u \frac{\partial u}{\partial \tau} + \sum_{n=1}^{\infty} D_n u_n (x) \sin n\omega\tau \qquad (2.5.4.1)$$

In Eq. 2.5.4.1 D_n is the dissipation parameter of the n-harmonic, $\vartheta = \vartheta' \frac{\omega_0}{V_0}$, and $u_n = \frac{2}{\pi} \int_0^{\pi} u(x,\tau') \sin n\omega\tau' d(\omega\tau')$ is its amplitude. If we assume as we did before that $D_n \sim n^b$ if the decay can be neglected the solution of Eq. 2.5.4.1 with $D_n \to 0$ becomes a simple Riemann wave. We are interested in the case when $D_n \to \infty$ for certain $n = k$ when the corresponding harmonics are suppressed. By developing in Fourier series:

$$u = \sum_{n=1}^{\infty} u_n (x) \sin n\omega\tau$$

we can rewrite Eq. 2.5.4.1 as

$$\frac{du_n}{dx} + D_n u_n = S_n$$

In Eq. 2.5.4.2 S_n is given by:

$$S_n = \frac{2}{\pi} \int_0^{\pi} \xi u \frac{\partial u}{\partial \tau'} \sin n\omega\tau' d(\omega\tau') \qquad (2.5.4.2)$$

If some D_n are large enough, so $u_n = S_n/D_n \to 0$ the decaying harmonics do not take part in the interaction.

In the following example two harmonics with frequencies ω and 2ω propagate together in a medium with $D_n \to \infty$ for $n \geq 3$ we have:

$$\frac{du_1}{dx} = \frac{\xi\omega}{2} u_1 u_2$$
$$\frac{du_2}{dx} = -\frac{\xi\omega}{2} u_1^2 \qquad (2.5.4.3)$$

On the other hand we have $u_1^2 + u_2^2 = u_0^2 = \text{constant}$. Therefore

$$\frac{u_2}{u_0} = \frac{u_{20} \cosh \chi x - u_0 \sinh \chi x}{u_0 \cosh \chi x - u_{20} \sinh \chi x} \text{ with } u_{20} = u_2(0) \text{ and } \chi = \xi\omega u_{20}/2 \qquad (2.5.4.4)$$

If $x \to \infty$ then $u_1 \to 0$ and $u_2 \to -u_1$ which means that the elastic energy is transferred from the first into the second harmonic. This cannot happen in a medium without dispersion and losses.

2.5.4.1 Parametric amplification

An interesting situation is the case of $u_1(0) \ll u_2(0)$. In the initial stage $u_2 = u_{20}$ and Eqs. 2.5.4.3 or 2.5.4.4 yields:

$$u_1 = u_{10}e^{\chi x} \tag{2.5.4.1.1}$$

This is called parametric amplification of the elastic wave of frequency ω by the 'pump' elastic wave of frequency 2ω. If $u_{20} > 0$ from Eq. 2.5.4.4 follows that u_1 will grow exponentially, while for $u_{20} < 0$ it will decay. This means that the parametric amplification or decay depends on the 'pumping' wave phase. The phenomenon of parametric amplification plays an important role in nonlinear optics. However, it can play also an important role in seismology. As we will see in the next Chapter it occurs in Rayleigh waves as well which is expected to have a great impact during earthquakes.

The above example shows that it is possible to generate elastic waves by elastic waves. This cannot happen without dispersion and dissipation. It is interesting to see what happens in a rectangular waveguide with travelling waves. Such a scenario is very possible to be observed with seismic waves in big rock slabs which walls create waveguide boundary conditions. We will assume that the propagation is along x-axis of the waveguide, while y- and z-axis form the cross-section of the waveguide with dimensions $a \times b$. A solution in the form of three interacting modes of frequencies $\omega_{1,2,3}$ will be sought (Naugolnykh and Ostrovsky 1998):

$$\varphi = \sum_s \sum_{m,n} f_{m,n}(y,z) A_s(x,t) e^{i(\omega_s t - k_s x)} \tag{2.5.4.1.2}$$

In Eq. 2.5.4.1.2 $s = 1,2,3$ and $f_{m,n}$ are the waveguide eigenfunctions determined by the boundary equation:

$$\frac{\partial^2 f_{m,n}}{\partial y^2} + \frac{\partial^2 f_{m,n}}{\partial z^2} + \chi_s^2 f_{m,n} = 0 \tag{2.5.4.1.3}$$

Equation 2.5.4.1.2 is a solution of the wave propagation equation in a homogeneous waveguide, where $\gamma = 1 + B/A$ with the ratio B/A characterizing the nonlinearity of the system and c is a constant (Andreev 1955; Aaonsen et al. 1984):

$$\nabla^2 \varphi - \frac{1}{c}\varphi_{tt} = \frac{1}{c^2}\frac{\partial}{\partial t}\left[(\nabla\varphi)^2 + \frac{\gamma-1}{2c^2}(\varphi_t)^2 \right] \tag{2.5.4.1.4}$$

The substitution of Eq. 2.5.4.1.2 into Eq. 2.5.4.1.4 with the appropriate eigenfunctions $f_{m,n}$, keeping only the resonant terms, and integration over the waveguide cross-section yields the equation of the mode amplitudes:

$$\frac{\partial A_m}{\partial t} + V_g \frac{\partial A_m}{\partial x} = G_m \qquad (2.5.4.1.5)$$

In Eq. 2.5.4.1.5 we have the group velocity of the mode V_g and a nonlinear term of mode amplitudes. For a process that does not evolve in time $\partial/\partial t = 0$ and in this case Eq. 2.5.4.1.5 leads to a triplet of equations:

$$i\frac{\partial A_1}{\partial x} = -B_1 A_2^* A_3$$

$$i\frac{\partial A_2}{\partial x} = -B_2 A_1^* A_3 \qquad (2.5.4.1.6)$$

$$i\frac{\partial A_1}{\partial x} = B_3 A_1 A_2$$

In the triplet of equations Eq. 2.5.4.1.6 B_1, B_2, B_3 are considered real.

Let us consider the interaction of two modes of frequencies ω and 2ω. With the assumption that there is a small frequency difference between the waves $\omega_2 = 2\omega_1 + \Delta\omega$ and $k_2 = 2k_1 + \Delta k$, where $\Delta\omega \ll \omega_1$ and $\Delta k \approx \Delta\omega dk_2/d\omega$ Eq. 2.5.4.1.6 becomes:

$$i\frac{\partial A_1}{\partial x} = -A_1^* A_2 B_1 e^{i(\Delta\omega t - \Delta kx)}$$

$$i\frac{\partial A_2}{\partial x} = A_1^2 B_2 e^{i(-\Delta\omega t + \Delta kx)} \qquad (2.5.4.1.7)$$

In terms of real amplitude and phase $A_{1,2} = a_{1,2} \exp(i\theta_{1,2})$ we get:

$$\frac{\partial a_1}{\partial x} = -a_1 a_2 \, \text{Im}\left[B_1 e^{i\theta}\right]$$

$$\frac{\partial a_2}{\partial x} = a_1^2 \, \text{Im}\left[B_2 e^{i\theta}\right] \qquad (2.5.4.1.8)$$

$$\frac{\partial \theta}{\partial x} = \Delta k + \text{Re}\left[B_1\left(a_2 - \frac{a_1^2}{a_2}\right)e^{i\theta}\right]$$

where $\theta = \theta_2 - 2\theta_1 - (\Delta k)x$.

We consider that the mode with amplitude a_2 acts a 'pump' and transfer energy to the weak mode with amplitude a_1 where $a_{10} \ll a_2$.

If $\Delta = 0$ (exact synchronism) the amplitude a_1 grows exponentially as we saw in Eq. 2.5.4.1.4:

$$a_1 = a_{10} e^q \tag{2.5.4.1.9}$$

In Eq. 2.5.4.1.9 $q = a_2 \, \text{Im}[B_1 e^{i\theta x}]$ and $K = e^q$ is the amplification factor. With a real B_1 a maximum equal to $B_1 a_2$ is attained at $\theta = \pi/2$.

The frequency band of amplification can be obtained by solving the system Eq. 2.5.4.1.8. If assuming large detuning between the modes, i.e., $\Delta k \gg B_1 a_2$ and setting $\theta = \Delta k x + \theta_0$ we get (Akhmanov and Khokhlov 1964):

$$a_1 = a_{10} \exp\left(B_1 a_2 \, \frac{\sin(x/2)\Delta k}{\Delta k / 2} \right) \tag{2.5.4.1.10}$$

From Eq. 2.5.4.1.10 we can conclude that a_1 is changing periodically along the x-axis with a period of $2\pi/k$. When the frequency ω_2 approaches a critical mode frequency where $k \to 0$ the amplification factor increases and the bandwidth narrows.

2.5.4.2 Elastic solitons in solids

In a nonlinear dispersive system, an initial disturbance can evolve into a solitary wave that retains its shape over a long distance. It has been found that when solitons collide with each other, they resume their initial wave forms and speeds.

Chiral rotational longitudinal waves or PR-waves and rotational shear waves SR-waves propagating faster in solid rocks and much slower in fractured media along tectonic faults have been observed to form rotational seismic solitons (Torres-Silva and Cabezas 2012). Because solitons can propagate without any loss of energy, these waves are extremely important carriers of seismic energy.

Solitons have been also observed experimentally in solids and theoretically analyzed (Hao and Maris 2001). The wave equation that governs the propagation of a finite-amplitude acoustic wave in a dispersive crystalline solid for a wave of wavelength sufficiently long that the effect of dispersion can be ignored so the standard nonlinear elasticity theory can be used can be written in the form:

$$\rho_0 \frac{\partial^2 u_\alpha}{\partial t^2} = \frac{\partial^2 u_\gamma}{\partial a_\beta \partial a_\delta} \left(A_{\alpha\beta\gamma\delta} + A_{\alpha\beta\gamma\delta\varepsilon\zeta} \frac{\partial u_\varepsilon}{\partial a_\zeta} \right) \tag{2.5.4.2.1}$$

where a_α is the Lagrangian coordinate in the direction α, u_α is the displacement in the direction α, ρ_0 is the unstrained density, and the coefficients $A_{\alpha\beta\gamma\delta}$ and $A_{\alpha\beta\gamma\delta\varepsilon\zeta}$ are defined by (Leibfried 1960):

$$A_{\alpha\beta\gamma\delta} = C_{\alpha\beta\gamma\delta}; \ A_{\alpha\beta\gamma\delta\varepsilon\zeta} = C_{\alpha\beta\gamma\delta\varepsilon\zeta} + C_{\alpha\beta\gamma\delta\zeta}\delta_{\gamma\varepsilon} + C_{\gamma\delta\beta\zeta}\delta_{\alpha\varepsilon} + C_{\varepsilon\zeta\beta\delta}\delta_{\alpha\gamma} \quad (2.5.4.2.2)$$

In Eq. 2.5.4.2.2 $C_{\alpha\beta\gamma\delta}$ and $C_{\alpha\beta\gamma\delta\varepsilon\zeta}$ are the second and third order elastic constants. Equation 2.5.4.2.1 is limited to quadratic nonlinearity of the displacement, third and higher orders are neglected.

The Lagrangian of the systems is:

$$L = \int_V \left(E_K - E_P\right)dv \quad (2.5.4.2.3)$$

The kinetic energy density is given by:

$$E_K = \frac{1}{2}\rho\left(\frac{\partial u_i}{\partial t}\right)^2 \quad (2.5.4.2.4)$$

The potential energy density is:

$$E_P = \frac{1}{2}C_{\alpha\beta\gamma\delta}\frac{\partial u_\alpha}{\partial x_\beta}\frac{\partial u_\gamma}{\partial x_\delta} + \frac{1}{6}A_{\alpha\beta\gamma\delta\varepsilon\zeta}\frac{\partial u_\alpha}{\partial x_\beta}\frac{\partial u_\gamma}{\partial x_\delta}\frac{\partial u_\varepsilon}{\partial x_\zeta} =$$
$$\frac{1}{2}c_{\alpha\beta\gamma\delta}S_{\alpha\beta}S_{\gamma\delta} + \frac{1}{6}c_{\alpha\beta\gamma\delta\varepsilon\zeta}S_{\alpha\beta}S_{\gamma\delta}S_{\varepsilon\zeta} \quad (2.5.4.2.5)$$

Equation 2.5.4.2.1 becomes much simpler if we consider a propagation of a longitudinal wave in a cubic symmetry crystal along one its principal directions. In this case we get:

$$\rho_0\frac{\partial^2 u}{\partial t^2} = \left(C_2 + C_3\frac{\partial u}{\partial a}\right)\frac{\partial^2 u}{\partial a^2} \quad (2.5.4.2.6)$$

C_2 and C_3 are combination of second and third order elastic constants.

If we want to include dispersion Eq. 2.5.4.2.6 becomes:

$$\rho_0\frac{\partial^2 u}{\partial t^2} = \left(C_2 + C_3\frac{\partial u}{\partial a}\right)\frac{\partial^2 u}{\partial a^2} + 2\vartheta\rho_0 V\frac{\partial^4 u}{\partial a^4}$$

where $\vartheta > 0$ is a constant. If we define the strain $\eta = \partial u/\partial a$ and differentiate Eq. 2.5.4.2.6 with respect to a we obtain:

$$\rho_0\frac{\partial^2 \eta}{\partial t^2} = C_2\frac{\partial^2 \eta}{\partial a^2} + C_3\frac{\partial}{\partial a}\left(\eta\frac{\partial \eta}{\partial a}\right) + 2\vartheta\rho_0 V\frac{\partial^4 \eta}{\partial a^4} \quad (2.5.4.2.7)$$

Equation 2.5.4.2.7 is similar to the Korteweg and De Vries (KdV) equation, which is an upgrade to Burgers equation for dispersive medium:

$$\frac{\partial \eta}{\partial t} = -B_1 \frac{\partial \eta}{\partial a} - B_2 \eta \frac{\partial \eta}{\partial a} - B_3 \frac{\partial^3 \eta}{\partial a^3} \qquad (2.5.4.2.8)$$

where B_1, B_2, B_3 are constants to be determined. By differentiating Eq. 2.5.4.2.8 with respect to time and use Eq. 2.5.4.2.8 again to replace the time-derivatives of the strain we get:

$$\frac{\partial^2 \eta}{\partial t^2} = B_1^2 \frac{\partial^2 \eta}{\partial a^2} + B_1 B_2 \eta \frac{\partial}{\partial a}\left(\eta \frac{\partial \eta}{\partial a}\right) + 2B_1 B_3 \frac{\partial^4 \eta}{\partial a^4} \qquad (2.5.4.2.9)$$

If we assume that $B_1 = (C_2/\rho_0)^{1/2} = V$, $B_2 = C_3/2\,\rho_0 B_1$, $B_3 = \vartheta$ then Eq. 2.5.4.2.9 becomes identical to Eq. 2.5.4.2.7 and the solutions of Eq. 2.5.4.2.8 are also solutions to Eq. 2.5.4.2.7. The KdV equation has two types of solutions: period solutions and solitons. The soliton solutions can be presented in the form (Whitham 1974):

$$\eta = \eta_0 \mathrm{sech}^2\left\{\left(\frac{B_2 \eta_0}{12 B_3}\right)^{1/2}(a - Ut)\right\} = \eta_0 \mathrm{sech}^2\left\{\left(\frac{C_3 \eta_0}{24 \rho_0 \vartheta V B_3}\right)^{1/2}(a - Ut)\right\}$$

$$U = V + \frac{B_2 \eta_0}{3} = V + \frac{C_3 \eta_0}{6\rho_0 c} \qquad (2.5.4.2.10)$$

In Eq. 2.5.4.2.10 U is the velocity of the soliton and η_0 is the maximum amplitude of the strain. Since $B_2 < 0$ and $B_3 > 0$ we have $\eta_0 < 0$, i.e., the pulse must be a compression, rather than rarefaction. The soliton velocity according to Eq. 2.5.4.2.10 is always larger than the velocity of the elastic wave. The magnitude of the dispersion does not affect the velocity, only its width.

The KdV solitons are stable and the initial pulse can evolve into one or more solitons.

The experimental setup for generation and detection of solitons is shown in Fig. 2.5.4.2.1 (Hao and Maris 2001). An Al film was deposited onto one side of the wafer to serve as a transducer for generating and detecting the acoustic pulses. To generate acoustic pulses light pulses from a Ti: Sapphire mode-locked laser are focused on the surface of the Al film with repetition time between pulses of 13.25 ns. The laser light absorbed in the film raises the film temperature. This sets up a thermal stress in the Al film, and a longitudinal acoustic pulse is launched into the sample. The returning elastic pulse results in a change in the optical reflectivity of the Al film; this change in reflectivity is detected by means of a time-delayed light pulse from the same laser—a probe pulse. The fractional change of reflectivity $\Delta R(t)/R$ is of the order of 10^{-5}. The travel time for the first acoustic pulse ranged from 70 ns for the Si sample to over 600 ns for SiO_2.

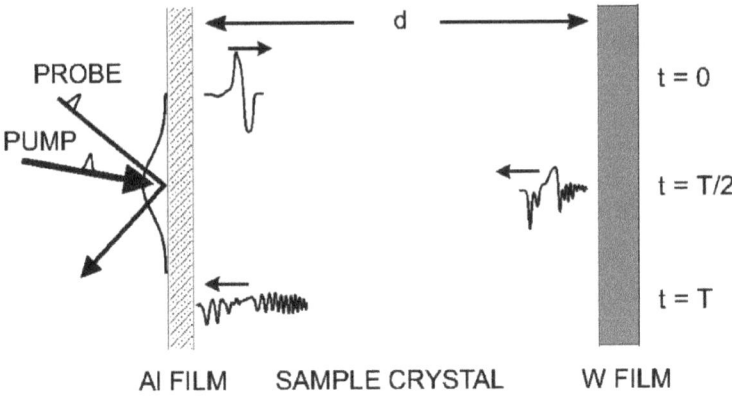

Fig. 2.5.4.2.1. Experimental setup for generation and detection of elastic soliton waves (Hao and Maris 2001).

When a strain pulse is generated in the Al film and propagates into the crystal, it will produce a soliton if the amplitude is sufficiently large. As noted in the previous section, the soliton must have a negative strain, i.e., it must correspond to a compression. If this strain pulse were to be reflected at a free surface of the sample, it would undergo a sign change and convert into a rarefaction pulse. The soliton would then be destroyed. To avoid this problem, a film of W is deposited onto the far side of the wafer. When the strain pulse is reflected at the interface between the wafer and the W film, the reflection coefficient is given by Eq. 2.5.3.1.3 where Z_1 is the W elastic impedance and Z_2 is the impedance of the sample. Since the elastic impedance of W is very large, the reflection coefficient is very large too and no change in sign of the strain occurs.

Fig. 2.5.4.2.2 shows the experimental results obtained in a sample of MgO and Fig. 2.5.4.2.3 shows the computer simulation (Hao and Maris 2001).

2.5.4.3 Stress-induced anisotropy in isotropic solid

Nonlinear elastic wave propagation is considered generally in two different approaches. The first assumes that the velocity of propagation of a finite-amplitude elastic wave depends on the strain. This means that an initially sinusoidal waveform will get distorted during the propagation in the absence of large dissipation of energy because the wave crests overtake the wave troughs leading ultimately to a shock wave (Zarembo and Krasil'nikov 1971; Hamilton 1986). For a P-wave propagating in an isotropic medium:

$$V_P(S) \approx V_P(S = 0)(1 - \beta S) \qquad (2.5.4.3.1)$$

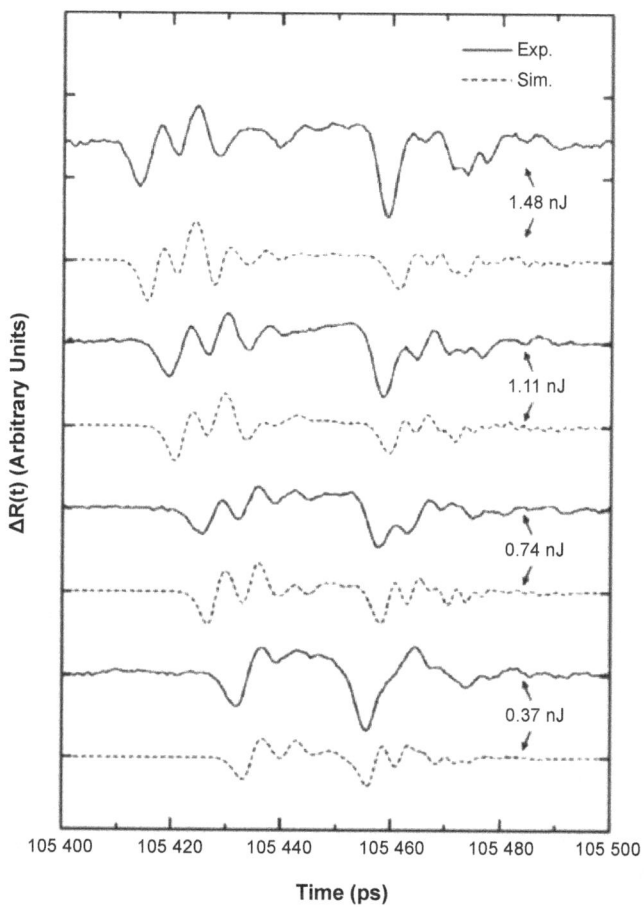

Fig. 2.5.4.2.2. First echo in MgO sample. The solid line is the experiment performed at 30 K. The dashed line is the computer simulation. Laser pulse energies are labeled (Hao and Maris 2001).

where $V_p(S)$ and $V_p(S = 0)$ are the linear and nonlinear velocity of the P-wave, S is the strain and β is the nonlinear elastic coefficient.

The second approach assumes a small wave perturbation superimposed on a static prestrain due to the presence of a static prestress (Pao 1984). Uniaxially and hydrostatically prestressed media have been analyzed (Thurston and Brugger 1964). In the case of uniaxial prestress the stress derivative of the wave modulus is given by:

$$\left(\frac{\partial \left(\rho_0 W^2 \right)}{\partial S} \right)_{S=0} = -\left(\vec{n} \cdot \vec{m} \right)^2 - 2wF - H \qquad (2.5.4.3.2)$$

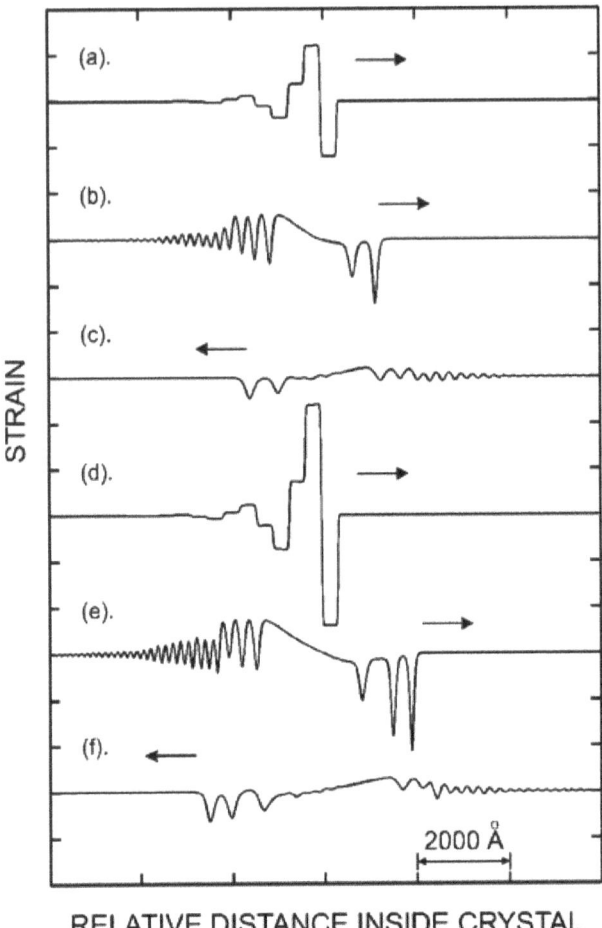

STRAIN

RELATIVE DISTANCE INSIDE CRYSTAL

Fig. 2.5.4.2.3. Computer simulation of the shape of strain pulses propagating in MgO. The upper part of the figure shows (a) the initial pulse entering the sample, (b) the pulse as it approaches the far side of the sample, and (c) as it returns to the Al film. In the lower part, (d), (e), and (f) show the propagation of a pulse of larger amplitude (Hao and Maris 2001).

where W is the 'natural' velocity, i.e., the length of the elastic path in the unstressed state divided by the wave travel time in the stressed state, $\bar{n}$ and $\bar{m}$ are the unit vectors of the direction of propagation in the unstressed state and of the uniaxial stress. The quantities w, F, H are given by:

$$w = C^S_{ijkl} n_i n_k p_j p_l$$

$$F = S^T_{ijkl} m_i m_j p_k p_l \tag{2.5.4.3.3}$$

$$H = S^T_{ijkl} C_{kltrsq} m_i m_j n_t n_s p_r p_q$$

In Eq. 2.5.4.3.3 C^S_{ijkl} and S^T_{ijkl} are the second-order isentropic and isothermal compliance tensors of the unstressed medium and p_j are the polarization vector components in the unstressed state.

In the case of hydrostatic priestess applied to the medium $(\vec{n} \cdot \vec{m})^2 = 1$ and w, F, H are given by:

$$w = C^S_{ijkl} n_i n_k p_j p_l$$

$$F = S^T_{iirs} p_r p_s$$

$$H = S^T_{iiuv} C_{uvprqs} n_p n_q p_r p_s \qquad (2.5.4.3.4)$$

The symmetry of the stressed medium is determined by the symmetry of the second- and third-order compliance tensors. Therefore the elastic symmetry is unchanged hydrostatic stress. In materials having $C_{ijklmn} \gg C_{ijkl}$ the physical nonlinear term H in Eqs. 2.5.4.3.2, 2.5.4.3.3, and 2.5.4.3.4 will be much more important than the geometrical nonlinear term $2wF$. This the case of strongly nonlinear materials such as rocks.

We will establish the relationships between nonlinearity and elastic parameters assuming hydrostatically induced prestress and transverse anisotropy induced by uniaxial stress (Johnson and Rasolofosaon 1996). First we want to express the variations of the elastic properties such as S-wave birefringence and P-wave anisotropy induced by a uniaxial stress applied to the medium. If we consider that medium of propagation is initially isotropic the only directional parameter for the wave propagation will be the angle θ between the direction of propagation of the wave and the direction of the uniaxial stress. The S-wave birefringence $B_S(\theta)$ can be expressed by:

$$B_S(\theta) = \frac{V_{S_1}(\theta) - V_{S_2}(\theta)}{V_{S_1}(\theta)} \qquad (2.5.4.3.5)$$

V_{S_1} and V_{S_2} are the velocities of the faster and slower S-waves, respectively. In this case these velocities are the velocities of SV- and SH-waves. If we assume that $S^2_{ij} C^2_{ijklmn} \ll (C^2_{ijkl})^2$, i.e., the stress is not too large, Eqs. 2.5.4.3.2, 2.5.4.3.3, and 2.5.4.3.4 yield the value of $B_S(\theta)$:

$$B_S(\theta) \approx -Sb_S \sin^2 \theta \qquad (2.5.4.3.6)$$

In Eq. 2.5.4.3.6 $b_S = (4\mu + n)/8\mu^2$ is an elastic parameter (see Eq. 2.5.2.28) called stress-induced S-birefringence coefficient. Maximum birefringence occurs when $\theta = \pi/2$ which corresponds to the case when the two S-waves propagate in the same direction that is perpendicular to the uniaxial stress direction.

In a similar way the P-wave anisotropy can be defined:

$$A_P(\theta) = \frac{V_{P\|} - V_P(\theta)}{V_{P\|}}$$
(2.5.4.3.7)

$V_{p\|}$ and $V_p(\theta)$ are the velocities along the stress direction (the fastest velocity) and in any other direction defined by the angle θ, respectively. If the prestress is not too large, as in the previous case of $B_S(\theta)$ the value of $A_p(\theta)$ is:

$$A_P(\theta) \approx -Sa_p \sin^2 \theta$$
(2.5.4.3.8)

Here again $a_p = (2\lambda + 5\mu + 2m)/[2\mu(\lambda + 2\mu)]$ is an elastic parameter called stress-induced P-wave anisotropy coefficient. The maximum P-wave velocity deviation is between the propagation direction parallel and perpendicular to the direction of the uniaxial stress.

Figures 2.5.4.3.1 and 2.5.4.3.2 show the experimental date for S-wave birefringence and P-wave anisotropy obtained in 'natural' and 'thermally cracked' sandstones (Zamora 1990).

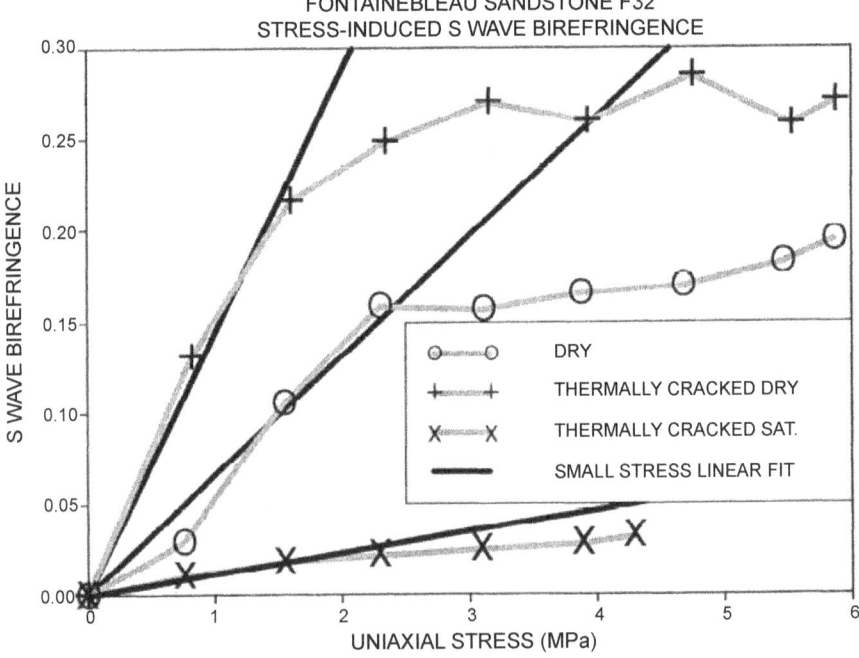

Fig. 2.5.4.3.1. S-wave birefringence as a function of uniaxial stress in 'natural' and 'thermally cracked' sandstones (Zamora 1990).

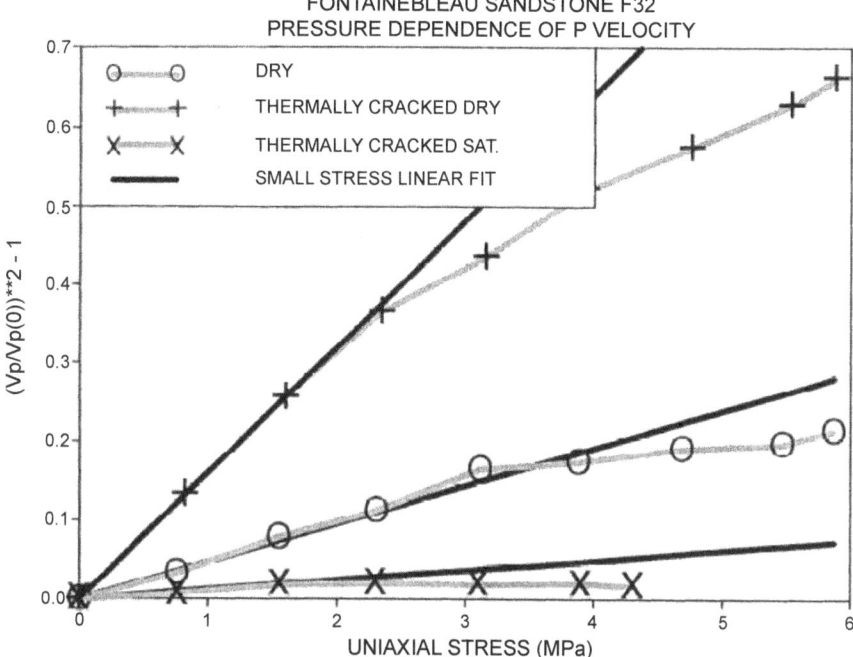

FONTAINEBLEAU SANDSTONE F32
PRESSURE DEPENDENCE OF P VELOCITY

Fig. 2.5.4.3.2. Relative variation of the P-wave velocity as a function of uniaxial stress in 'natural' and 'thermally cracked' sandstones (Zamora 1990).

2.5.5 Wave propagation in heterogeneous media

Dispersive wave propagation in heterogeneous solids is caused mainly by scattering of elastic waves. This situation is the closest to the propagation of seismic waves. The underlying physics governing finite-amplitude wave propagation in heterogeneous media will be discussed in this section. Of particular interest are those aspects of elastic heterogeneous solids that lead to structured steady shock waves. Theoretical studies based on a quasi-harmonic normal mode representation of the scattered wave energy have been reported. Elastic waves scattered in a heterogeneous solid during passage of a directed shock wave or transient pulse are, in principal, the same in nature as the lattice waves which characterize the thermal state of a solid (Grady 1997). They differ only in the wave frequencies which are substantially higher in thermal waves than microstructural elastic waves.

Finite-amplitude compression waves in nonlinear solids become steeper as the wave propagates. Wave dispersion can counter the steepening process leading to structured steady shock waves. Heterogeneities within a solid body contribute to wave dispersion and can account for structured steady shock waves. Typical shock wave profile is shown in Fig. 2.5.5.1.

Experimental study of shock wave propagation in periodically layered composites has been reported (Zhuang et al. 2003). In heterogeneous media, scattering due to interfaces/microstructure between dissimilar materials could

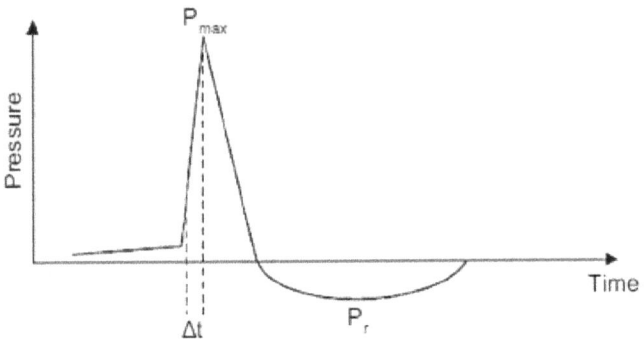

P_{max} – pressure maximum, P_r – negative peak pressure, Δt – pressure rise time

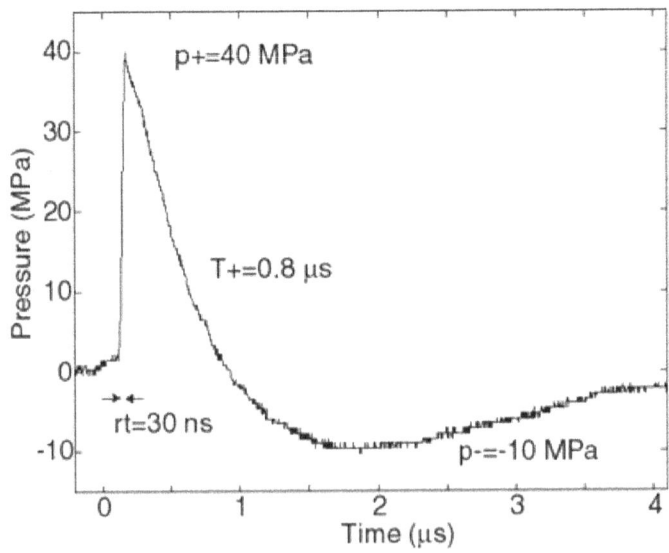

Fig. 2.5.5.1. Typical profile of a shock pulse.

play an important role in shock wave dissipation and dispersion. Experimental results (obtained using velocity interferometer and stress gage) show that these periodically layered composites can support steady structured shock waves. Due to interface scattering, the effective shock viscosity increases with the increase of interface impedance mismatch, and decreases with the increase of interface density (interface area per unit volume) and loading amplitude. The strain rate within the shock front is found to be roughly proportional to the square of the shock stress. This indicates that layered composites have much larger shock viscosity due to the interface/microstructure scattering in comparison with the increase of shock strain rate by the fourth power of the shock stress for homogeneous metals (Grady 1998). Experimental results also show that due to the scattering effects, shock propagation in the layered composites is dramatically slowed down and the shock speed in composites can be lower than that either of its components.

So far our discussions have been limited to quadratic nonlinearity only, described in the classic terms of the five-constant model. Here we want to extend the discussion by involving cubic nonlinearity. Cubic nonlinearity has barely been studied because of the much higher complexity of the governing equations of state. However, a logic question to ask is whether considering cubic nonlinearity is relevant in the case of seismic waves and is it worth it to extend the complexity of already complicated matter to quadratic nonlinearity? The cubic terms in the governing equations usually are much smaller in magnitude than the quadratic ones causing nonlinear waveform distortions in materials used in physical acoustics and acoustoelectronics. However, the presence of strong dispersion, as it is the case of heterogeneous media, or specific nonlinearity mechanisms as thermal phenomena may change the situation drastically (Naugolnykh and Ostrovsky 1998).

For many materials including the Earth's lithosphere much more complicated nonlinear characteristics have been observed including anomalously strong nonlinearity. The exact reasons for such strong nonlinearity are not well understood, but many experimental results point to heterogeneous structure of materials such as dislocations, micro-cracks, grains, fluids-filled pore space of rocks, hydrocarbon-reservoir rocks, and scattered rock domains of high stress serving as internal sources of seismic energy 'pumping' passing P- and S-waves. Seismic waves in earth materials are subject to attenuation and dispersion in a broad range of frequencies and scales from free oscillations of the entire earth to ultrasound in small rock samples (Aki and Richards 1980). The third-order nonlinear constants of rocks have been found to much larger than the second-order nonlinear constants (Johnson and Rasolofosaon 1996). Contrary to homogeneous solids with constant level of nonlinearity, rocks exhibit wide range of nonlinearity—from weak to very strong—increasing always their elastic moduli with pressure. As a result of this moduli variation the P-wave stress-induced anisotropy and S-wave stress-induced birefringence can be large.

Including cubic nonlinearity changes significantly the elastic fields with phenomena such as self-focusing, self-transparency, stimulated scattering, amplitude-dependent friction, etc. of elastic beams.

Equation 2.5.1.3 shows that the stress $T = T(S)$ is formulated as power series of the strain S. For media with anomalous nonlinearity the stress-strain relation $T = T(S)$ can be complicated. Figure 1.1.2 shows a reversible hysteresis of the stress-strain relation, however, there are many cases where the stress-strain hysteresis is not reversible as shown in Fig. 2.5.5.2.

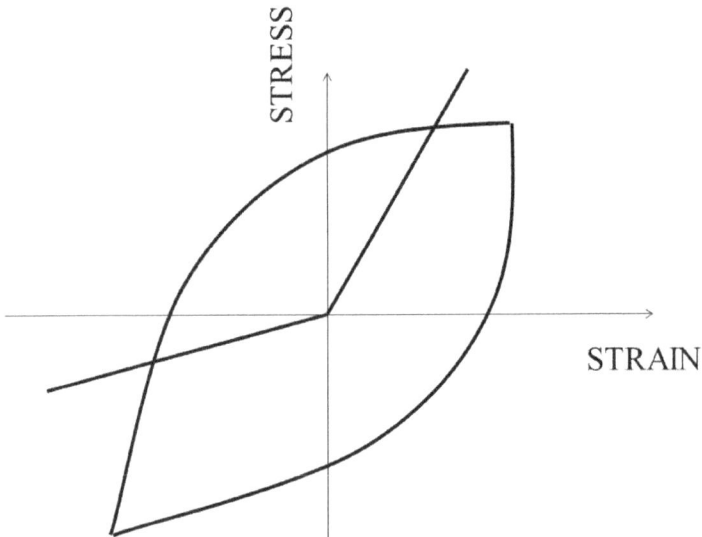

Fig. 2.5.5.2. Irreversible stress-strain hysteresis (red line) and bilinear stress-strain relation.

In geology a phenomenon called bimodular elasticity has been observed in rocks when elastic moduli under compression can become smaller or larger than under extension (Johnson and Rasolofosaon 1996). This means that a propagating P-wave causing periodic compressions and extensions will modulate its own velocity if it travels through a bimodular rock adding to the already existing dispersion of the medium.

3

Surface Elastic Wave Propagation

Most of the seismic energy during earthquakes is carried by body elastic waves in the Earth's lithosphere. Usually body waves have little impact to man-made constructions on the ground. Body elastic waves could have some effect only on deep underground constructions such as tunnels, dams, and gas or oil lines. Their impact is mostly through their conversions into surface elastic waves. Since most of buildings, highways, power stations, bridges and other man-made constructions are located above the ground surfaces elastic waves generated by bulk elastic waves are of greatest importance. There are various types of surface elastic waves that originate from body waves. In this section we will discuss how they are excited and how they propagate as well as their linear and nonlinear properties.

One of the most important questions is how to determine the locations where surface seismic waves are created during an earthquake. In seismology the point on the surface above the hypocenter is called epicenter of the earthquake (Fig. 3.1). This point will be the first to be reached by the P-waves coming from the hypocenter because they are the fastest ones. The P-waves will be followed by slower S-waves. Other waves will reach the Earth's surface some time later at points located around the epicenter. The P- and S-waves reaching the epicenter are coming almost normally to the free surface. They will be either reflected back into the bulk without forming any new waves capable to propagate on the surface or depending on the roughness of the surface they could be scattered and converted into Rayleigh surface waves. Waves coming to the surface under various angles will also get scattered or reflected back after undergoing mode conversions or depending on their polarization they will form surface elastic waves that could propagate farther on the surface eventually reaching cities. These waves will actually 'participate' in the earthquake. The point of their formation on the Earth's surface is not

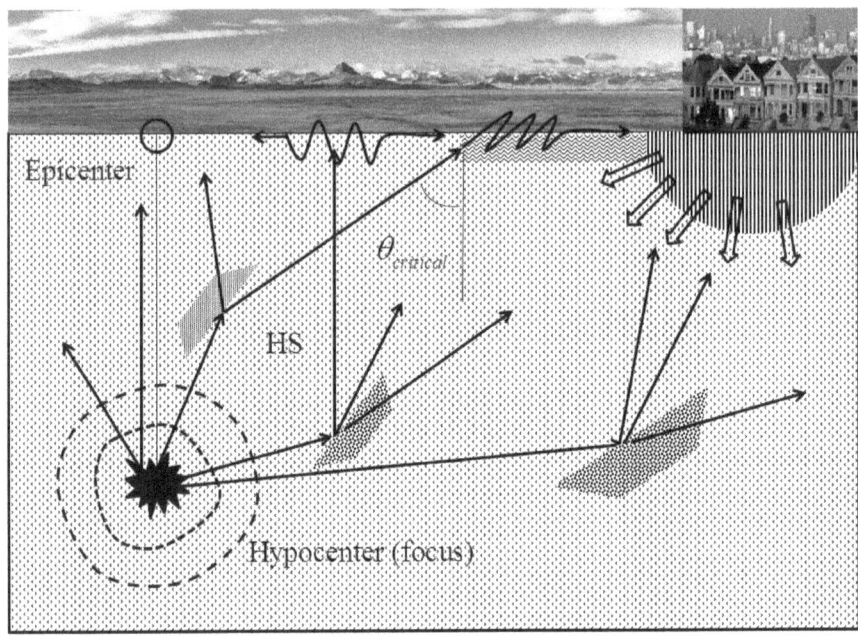

Fig. 3.1. Schematic of an earthquake focus, epicenter, and seismic wave propagation. A shied around the city can deflect or absorb the upcoming elastic waves.

necessarily located in the epicenter but somewhere else, probably far away from it. Many such points could exist on the Earth's surface where surface elastic waves can be generated in appropriate conditions and continue their propagation on the surface. Many waves that cannot meet specific boundary conditions will return back into the crust as body waves or will be attenuated.

It is relevant to ask the question where exactly the seismic waves that are registered by the seismological equipment are coming from and how is the epicenter of an earthquake determined? Usually to determine the location of the epicenter the measurements of at least three seismologic stations located far away of each other are registered and the location of the epicenter is calculated from the time intervals between the seismic waves arriving at the seismic stations knowing that P-waves are faster than S-waves and S-waves are faster than the surface acoustic waves (LR, LQ, or SAW). Are these waves primary waves coming from the hypocenter or are they secondary waves that have been generated by mode-conversions during multiple reflections and refractions by the rock layers forming the heterogeneous crust? Self-modulation of carrier waves of the diffuse elastic field (coda waves) as well as local amplification of seismic waves due to nonlinear phenomena or specific geological properties of soil could affect strongly the shape of seismograms recorded at different points. A typical example are the three seismographs shown in Fig. 1.1.1.7 that have been recorded during the same earthquake but at different seismic stations from the epicenter of the 1999 Kansas City collapse event. These seismograms are a clear demonstration that the registered seismic waves are not the same routine P-, S- or LR-waves that have travelled different distances.

Seismogram recordings allow determining the moment magnitude of an earthquake on the Gutenberg-Richter scale as well as particle velocity and acceleration at the seismometer's location. However, from the recordings it is not possible to define the type of seismic waves that have reached a city or their characteristics. This is the main reason that both Gutenberg-Richter and Mercalli scales are used together. Knowing how elastic waves propagate it is clear that the detected waves are rarely coming directly from the focus of the earthquake. In most of cases they originate from the same initial seismic disturbance but have been modified during their propagation in the heterogeneous nonlinear rock. This is especially true for surface seismic waves which always originate from body waves reaching the Earth's surface under specific conditions. In this aspect the practical importance of the exact location of the epicenter of an earthquake that is located above the hypocenter is of lesser importance than the depth of the focus (the distance between the hypocenter and the epicenter) which is an important parameter for the characterization of the earthquake. Usually the high-amplitude primary waves are not a direct threat to cities unless the city happens to be located close to the epicenter, but they are important source of surface seismic waves which have the strongest impact to man-made constructions.

In Sections 3.2, 3.3, and 3.4 three different types of surface elastic waves will be analyzed that are often generated during earthquakes. The damage caused to building by these waves depends on the seismic energy carried by them but mostly it depends on the polarization of the waves. We remind that the polarization is the direction in which ground masses get displaced. While vertical displacement normal to the ground surface could be damaging much more damaging are waves with horizontal sidewinding-snake type displacement SH-wave.

3.1 Linear surface elastic wave propagation

Surface elastic waves propagating on the planar surface of a homogeneous, linear, elastic half-space are nondispersive. Two types of dispersion can exist: (1) material dispersion associated with attenuation, scattering and diffraction of the waves due to the structure of the medium of propagation, and (2) geometrical dispersion in layered structures and elastic waveguides which is due to wave interference. Usually material dispersion is weaker that geometrical dispersion. Material substrates used in physical acoustics for generation and propagation of surface acoustic waves (SAW) usually are nondispersive.

If the material is nonlinear and nondispersive an initially sinusoidal surface elastic wave creates higher harmonics which in the absence of dispersion can grow indefinitely interacting with each other and also with the fundamental mode. Higher harmonics of Rayleigh waves have been observed experimentally since 1960s (Mayer 1995). A derivation of coupled equations for slow variations of the envelopes of a fundamental Rayleigh wave and its higher harmonics had been given as early as 1973 (Reutov 1973). In nonlinear dispersive material the growth of higher harmonics and their interactions are inhibited by the dispersion.

Equation 2.1.1 admits solutions that correspond to inhomogeneous dispersive plane waves satisfying the boundary conditions of surface waves propagating on the surface (x_1, x_2) of a solid occupying the space $x_3 \leq 0$ (Fig. 3.1.1). The displacement can be presented as (Farnell 1970; Farnell and Adler 1972):

$$\vec{u}_i = \alpha_i e^{i(\vec{k}\cdot\vec{x}-V_R t)} \tag{3.1.1}$$

The components $\vec{k} = (k_1, k_2, k_3)$ in the (x_1, x_2)-plane are real for k_1 and k_2 while k_3 may be complex. For a plane wave propagating along the x_1-axis $k_1 \neq 0$, $k_2 = 0$ and $k_3 = i\xi$ is complex.

$$\vec{k} = \left(k_1, 0, i\xi\right) \tag{3.1.2}$$

Equation 3.1.1 becomes:

$$u_i = \alpha_i e^{i\xi k x_3} e^{i(k_1 x_1 - \omega t)} \tag{3.1.3}$$

Equation 2.1.2 becomes:

$$\rho V_R^2 \alpha_i = c_{ijkl} l_j l_l \alpha_k \tag{3.1.4}$$

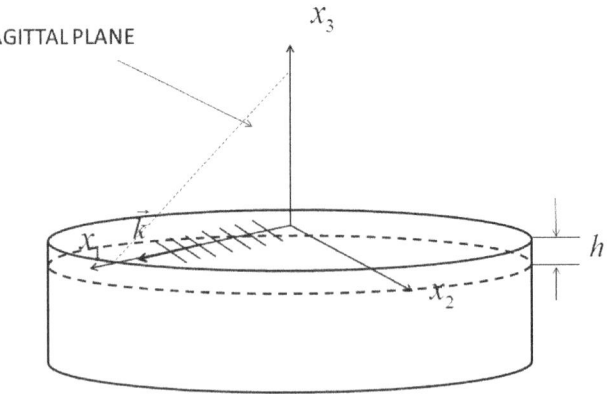

Fig. 3.1.1. Surface elastic wave propagating along x-axis.

l_j is the unit vector of the axes x_j. Christoffel's equation Eq. 2.1.6 is now:

$$\left| \Gamma_{il}(\xi) - \rho V_R^2 \delta_{il} \right| = 0 \text{ with } \delta_{il} = 0 \text{ if } i \neq l \text{ and } \delta_{il} = 1 \text{ if } i = l \qquad (3.1.5)$$

Also we have:

$$\Gamma_{ik} = c_{ijkl} l_j l_l = \Gamma_{ki} \qquad (3.1.6)$$

(With substituting Eq. 3.1.3 into Eq. 3.1.5 we get:

$$
\begin{aligned}
\Gamma_{11} &= c_{11} + 2c_{15}\xi + c_{55}\xi^2 \\
\Gamma_{22} &= c_{66} + 2c_{46}\xi + c_{44}\xi^2 \\
\Gamma_{33} &= c_{55} + 2c_{35}\xi + c_{33}\xi^2 \\
\Gamma_{12} &= c_{16} + \left(c_{14} + c_{56}\right)\xi + c_{45}\xi^2 \\
\Gamma_{13} &= c_{15} + \left(c_{13} + c_{55}\right)\xi + c_{35}\xi^2 \\
\Gamma_{23} &= c_{56} + \left(c_{36} + c_{45}\right)\xi + c_{34}\xi^2
\end{aligned}
\qquad (3.1.7)
$$

Since we deal with surface elastic waves all displacement components should become 0 with $x_3 \rightarrow -\infty$, i.e., only solutions with Im $\xi < 0$ are allowed in Eq. 3.1.3:

$$u_i = \sum_{s=1}^{3} C_s \alpha_i^{(s)} e^{i(k_x x_1 - V_R t)} = \sum_{s=1}^{3} C_s \alpha_i^{(s)} e^{i\xi_3 x_3} e^{i(k_1 x_1 - \omega_R t)} \qquad (3.1.8)$$

The coefficients are chosen to satisfy the free half space boundary conditions:

$$T_{i3} = 0 \text{ at } x_3 = 0 \qquad (3.1.9)$$

From Eq. 1.5.1.6 we have:

$$c_{i3kl} \frac{\partial u_k}{\partial x_l} \text{ at } x_3 = 0 \qquad (3.1.10)$$

Equation 3.1.8 becomes:

$$c_{i3kl} \sum_{s=1}^{3} C_s \alpha_k^{(s)} l_l^{(s)} = 0 \qquad (3.1.11)$$

In matrix form Eq. 3.1.11 is:

$$
\begin{pmatrix}
\Gamma_{11} - \rho V^2 & 0 & \Gamma_{13} \\
0 & \Gamma_{22} - \rho V^2 & 0 \\
\Gamma_{12} & 0 & \Gamma_{33} - \rho V^2
\end{pmatrix}
\begin{pmatrix}
\alpha_1 \\
\alpha_2 \\
\alpha_3
\end{pmatrix}
= 0
\qquad (3.1.11')
$$

From Eq. 3.1.4 and Eq. 3.1.11 the parameters V_R, ξ_s, C_s, and $\alpha_i^{(s)}$ can be calculated. For an isotropic substrate Eq. 3.1.7 becomes:

$$\Gamma_{11} = c_{11} + c_{44}\xi^2$$
$$\Gamma_{22} = c_{44}\left(1 + \xi^2\right)$$
$$\Gamma_{33} = c_{44} + c_{11}\xi^2$$
$$\Gamma_{12} = 0 \tag{3.1.12}$$
$$\Gamma_{13} = \left(c_{11} - c_{44}\right)/\xi$$
$$\Gamma_{23} = 0$$

With $\xi = 0$ we find the solutions for bulk elastic P- and S-waves with $V_P = (c_{11}/\rho)^{1/2}$ and $V_S = (c_{44}/\rho)^{1/2}$.

With the boundary conditions the Christoffel equation Eq. 2.1.6 allows two real solutions of wave propagations in the plane (x_1, x_2)—one corresponding to a surface elastic wave with polarization in the sagittal plane (Rayleigh-like modes) and one corresponding to a surface elastic wave with horizontal polarization in the plane (x_1, x_2)—Love modes. The soliton corresponding to a wave that propagates along x_3-axis decays fast in the direction $- x_3$. Love waves are generated if the incident wave is a body SH-wave that is incident to the layer-substrate interface at certain critical angle of incidence such that the layer acts as a waveguide. Love waves are guided elastic waves that can propagate on the interface between free spaces and a substrate material on top of which there is a plate of another material which elastic properties are different from those of the substrate. The layer waveguide geometry makes them dispersive.

3.2 Rayleigh waves

Is it possible to generate some other type of surface elastic waves without the presence of the layer serving as a waveguide that we discussed in the case of Love waves in Section 3.1? Rayleigh in 1885 demonstrated that this is possible; however, the surface elastic wave is not linearly polarized as in the case of Love wave. In order to satisfy the boundary condition of a free surface with zero strain the polarization of the Rayleigh's wave should be formed by coupled P-wave and SV-wave similar to Lamb waves.

In Section 2.2.3 we discussed the case of a free plate with an SH-wave propagating and the existence of a cutoff frequency beyond which the wave stop propagating as well as SV-wave propagation and Lamb waves. Wave dispersion due to geometrical configurations such as half space, free plate, plate on top of a half space in contact with a fluid, etc. has many similarities with waveguide propagation in optics. The dispersion curves for the fundamental symmetric and antisymmetric Lamb modes in Fig. 2.2.3.2 show that for large values of ωb both symmetric and antisymmetric Lamb waves reach a constant value $V_R < V_S$, V_P. The transverse components k_{tP} and k_{tS} in Eq. 2.2.3.2 become imaginary when $\omega b \rightarrow \infty$ and both symmetric and antisymmetric Lamb waves become degenerated with same velocities of propagation. The modes are propagating only close to the boundary without elastic waves inside the plate. Since the thickness of the plate does not matter in this case the solutions are valid for a free half space. These solutions are called Rayleigh surface waves. Lamb waves propagate in a waveguide formed by a free-boundary plate while Rayleigh waves propagate on the surface without need of the presence of waveguide conditions. The velocities of propagation of both P- wave and SV-wave approaches a constant value $V_R < V_{SV}$, V_P. Similarly to Lamb's waves Rayleigh waves are couples waves of P- and SV-waves that propagate together. This means that on the free surface the particles' displacements follow a complex trajectory in the direction of propagation and perpendicularly to it. In the case of an isotropic medium of propagation the polarization of a Rayleigh wave is elliptical—'ground roll' with one longitudinal component and one shear component polarized perpendicularly to the free surface. The velocity of propagation of a Rayleigh wave is about 10% less than the velocity of propagation of the bulk shear wave with same polarization (see Fig. 3.2.2). We recall that in an isotropic medium of propagation one fast P-wave and two slower S-waves can propagate. The S-waves have the same velocity—they are degenerated. If the wave is considered to be plane wave then the polarizations of the two S-waves are orthogonal. If the medium of propagation is anisotropic Rayleigh waves can propagate on the free surface with velocity that is smaller than the velocity of the slowest bulk S-wave. That means the Rayleigh wave cannot phase match any bulk wave. Below the free surface the Rayleigh wave amplitude decays exponentially. Figure 3.2.1 illustrates the propagation of a Rayleigh wave.

It is interesting to compare Rayleigh waves to Lamb waves. Lamb waves are reflected from the two boundaries of the free plate, while Rayleigh waves have only one reflecting boundary. In this case there are no incident waves coming from the inside of the plate, but the reflected waves have to exist. This requires the reflection coefficients in Eq. 2.2.3.1 to become infinite. The transverse resonance condition for Rayleigh wave is the denominator Eq. 2.2.2.1 to become equal to 0:

$$\sin 2\theta_R^{SV} \sin 2\theta_I^P + \left(V_P / V_S\right)^2 \cos^2 2\theta_R^{SV} = 0 \qquad (3.2.1)$$

Also since both waves are evanescent in the depth of the plate the transverse components of the wave numbers are imaginary: $k_{tS} = i\alpha_{tSV}$ and $k_{tP} = i\alpha_{tP}$. Also $\sin 2\theta_R^{SV} = \dfrac{\beta_R V_{SV}}{\omega}$, $\sin 2\theta_I^P = \dfrac{\beta_R V_P}{\omega}$, $\cos 2\theta_R^{SV} = -i\dfrac{\alpha_{tSV} V_{SV}}{\omega}$, and $\cos 2\theta_R^P = -i\dfrac{\alpha_{tP} V_P}{\omega}$. The Eq. 3.2.1 becomes (Auld 1973):

$$\left[2\beta_R^2 - \left(\frac{\omega}{V_{SV}}\right)^2\right] - 16\beta_R^4\left[2\beta_R^2 - \left(\frac{\omega}{V_{SV}}\right)^2\right]\left[2\beta_R^2 - \left(\frac{\omega}{V_P}\right)^2\right] = 0 \qquad (3.2.2)$$

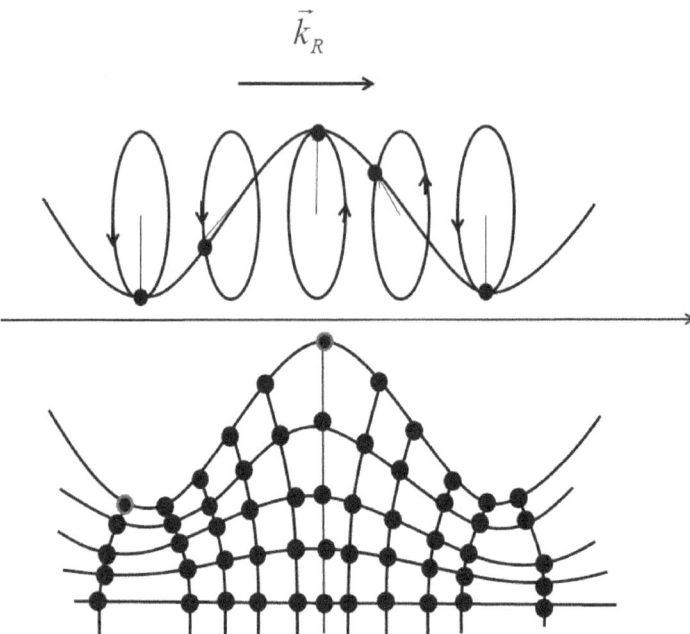

Fig. 3.2.1. Rayleigh wave is composed by one P-wave and one SV-wave elliptically polarized.

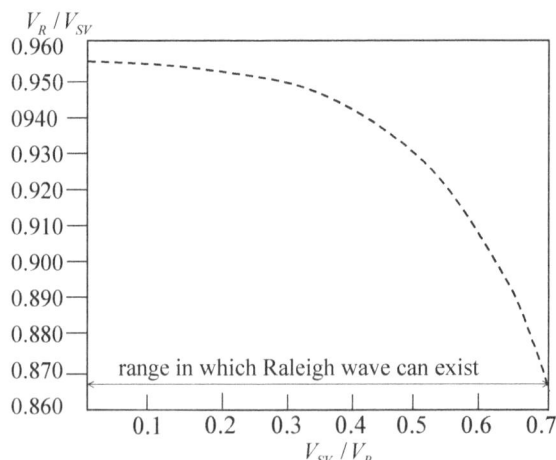

Fig. 3.2.2. Rayleigh wave velocity in isotropic medium as a function of the body SV-wave and the body P-wave (Auld 1973).

Lamb waves are propagating together but they are not fully coupled. Rayleigh waves are. In the case of Lamb waves propagating in a free plate two waves propagate at different velocities—one P-wave and one SV-wave. What exactly 'coupled' waves mean? This means that the two waves cannot exist independently without each other; they propagate together, and exchange energy between them. Each time when a SV-mode gets reflected by one of the boundaries of the plate it splits its energy between one new P-wave and one new SV-wave. Still these waves can propagate pretty independently—they are not fully coupled. There is no such thing in the case of Love waves because an SH-wave reflects into an SH-wave without mode conversion. In the case of Rayleigh waves an incident P-wave or SV-wave coming from the bulk hits the free boundary under a critical angle as a result of which a P-component along the free boundary is created and a S-component under certain angle to the free boundary. The ground particles start a roll motion following elliptical paths that are pushed further by the P-wave. The situation is similar to the propagation of ocean waves that are generated by the wind component parallel to the ocean surface.

This type of surface elastic waves is the most common in earthquakes. To form and propagate they do not need a waveguide structure on the boundary surface as it is the case of Love waves. The only request to generate Rayleigh wave on the free surface is to get one P- and one SV-wave coming on the surface from the bulk seismic field under the critical angle. The problem is that it is difficult to locate exactly the point where they will pop up on the ground surface, basically their 'epicenter'.

3.2.1 Dispersion of Rayleigh waves

For the isotropic solid both the wave equation and the boundary conditions give two independent types of solutions: surface elastic waves with horizontal displacement (polarization) called Love modes and surface elastic waves with displacement in the sagittal plane called Rayleigh waves (Farnell and Adler 1972). Raleigh waves are not dispersive unless they propagate in a layered structure as shown in Fig. 3.1.1. A film of thickness h and different elastic properties is deposited on top of the surface where Raleigh waves propagate. We will consider that only sagittal-plane displacements are involved. In this case the boundary conditions are more complex and solving the equation of motion requires numerical processing. If the layer thickness is small ($kh \ll 1$) some physics assessment can be done. In the limit case of $k \to 0$ the solution is nondispersive Raleigh waves on an isotropic unlayered half space (Stonley 1955; Victorov 1967). If $kh \to 0$ it has been found that

$$4\left[1-\left(\hat{V}_S / \hat{V}_P\right)^2\right]\hat{V}_S^2 > V_R^2\left[1+\left(\frac{1-\left(V_R / V_P\right)^2}{1-\left(V_R / V_S\right)^2}\right)^{1/2}\right] \tag{3.2.1.1}$$

where $\hat{V}_S = \sqrt{2}V_S$ and $\hat{V}_P = V_P / \sqrt{2}$.

Equation 3.2.1.1 can be simplified to

$$\hat{V}_S / V_S > \left[\frac{1-\left(V_S/V_P\right)^2}{\left(\hat{V}_S/\hat{V}_P\right)^2} \right]^{1/2}$$

(3.2.1.2)

The extreme limits of the right-hand side of Eq. 3.2.1.2 are represented by the lines $\hat{V}_S = \sqrt{2}V_S$ and $\hat{V}_S = V_S / \sqrt{2}$ (Fig. 3.2.1.1) which means that for layer-substrate combinations lying above the $\hat{V}_S = \sqrt{2}V_S$ line the layer is said to 'stiffen' the substrate because of its presence the SAW velocity increases above that of Raleigh velocity on the substrate, whereas for material combinations below $\hat{V}_S = V_S / \sqrt{2}$ line the layer is said to 'load' the substrate because the velocity of the free-surface Raleigh wave on the substrate decreases because of the presence of the layer (Fig. 3.2.1.1).

Two cases have been studied: $\hat{V}_S > V_S$ and $\hat{V}_S < V_S$ (Farnell and Adler 1972). The phase and group velocities have been calculated for both cases of Si on ZnO where $\hat{V}_S > V_S$ is valid and ZnO on Si where $\hat{V}_S < V_S$ is valid as function of increasing kh. In the first case the phase velocity starts from the point $kh = 0$ (corresponding to Raleigh velocity V_R in the substrate) and increase with increasing values of kh until it reaches the S-wave velocity V_S in the substrate. In the second case the phase velocity starts from V_R in the substrate at $kh = 0$ and decreases with kh increasing until reaching the values of $\hat{V}_S (> \hat{V}_R)$ in the layer tending asymptotically with $kh \gg 1$ to V_R which is the Raleigh wave velocity on the free surface of the layer material.

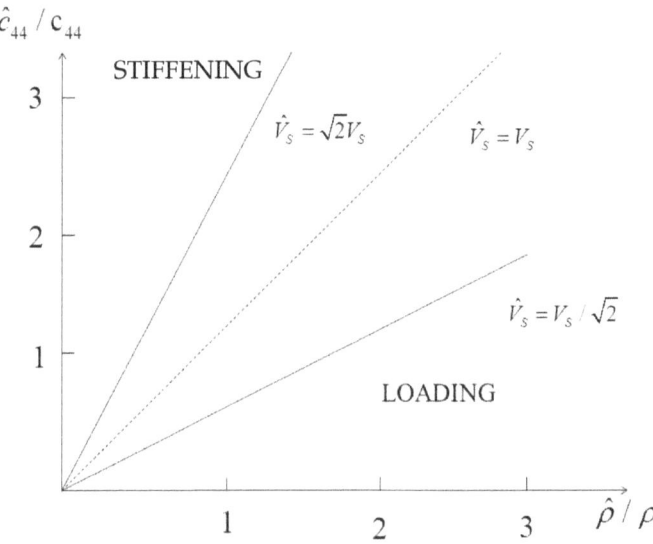

Fig. 3.2.1.1. Stiffening' and 'loading' for isotropic materials combinations (Farnell and Adler 1972).

3.3 Love waves

As mentioned above another possible solution of the wave equation and boundary conditions is a surface wave with displacement in the horizontal plane perpendicular to the sagittal plane. That could happen if the incident wave is a body SH-wave is incident to the layer-substrate interface at certain critical angle of incidence such that the layer acts as a waveguide. The velocity of propagation of the wave varies with frequency and its value is somewhere between the shear wave velocity in the thin film and the velocity in the substrate. This dispersion is typical for this type of surface acoustic waves called Love waves that has a great importance for seismic phenomena.

Love waves are guided elastic waves that can propagate on the interface between free spaces and a substrate material on top of which there is a plate of another material which elastic properties are different from those of the substrate (Fig. 3.3.1). They are polarized in the plane of the waveguide and their amplitude attenuates exponentially in the substrate. In the Section 2.2.1 it was mentioned that for an S-wave with a horizontal polarization SH (medium's particles displacement is parallel to the boundary) if $V_2/V_1 > 1$ there is a critical angle of incidence beyond which total reflection occurs. When medium 1 is a plate the SH-wave remain confined in that space as a result of total reflection. Medium 1 is acting as a waveguide. This Love wave belongs to the substrate-layer set. Love waves are dispersive with velocity of propagation dependent on frequency. To be generated the velocity of propagation of a horizontally polarized SH-wave should be slower in the layer than in the substrate. Love waves are polarized horizontally as is the S-wave that causes them. The existence of the plate firmly bonded to the substrate is mandatory request for the formation of Love waves. Following Auld's transverse resonance analysis (Auld 1978) for SH-waves propagating in the plate we have the particle velocities for incident, transmitted, and refracted waves:

$$v_{xI}^{(1)} = A e^{-i\left(-k_{iSH}^{(1)}y + \beta z\right)}$$

$$v_{xR}^{(1)} = B^{(1)} e^{-i\left(k_{iSH}^{(1)}y + \beta z\right)} \tag{3.3.1}$$

$$v_{xT}^{(2)} = B^{(2)} e^{-i\left(-k_{iSH}^{(2)}y + \beta z\right)}$$

Since $R_{SH}^{SH} = 1$ at the free boundary $y = b/2$ (see Section 2.3.2):

$$R_{SH}^{SH} = \frac{v_{xI}^{(1)}}{v_{xR}^{(1)}} = \frac{A e^{ik_{iSH}^{(1)}b/2}}{B^{(1)} e^{ik_{iSH}^{(1)}b/2}} = 1 \tag{3.3.2}$$

At the boundary $y = -b/2$ the reflection and transmission coefficients Eq. 2.3.1.2 are:

$$R_{SH}^{SH} = \frac{Z_1 \cos\theta_I^{SH} - Z_2 \cos\theta_T^{SH}}{Z_1 \cos\theta_I^{SH} + Z_2 \cos\theta_T^{SH}} = \frac{v_{xR}^{(1)}}{v_{xI}^{(1)}} = \frac{B^{(1)} e^{ik_{iSH}^{(1)}b/2}}{A e^{-ik_{iSH}^{(1)}b/2}} \tag{3.3.3}$$

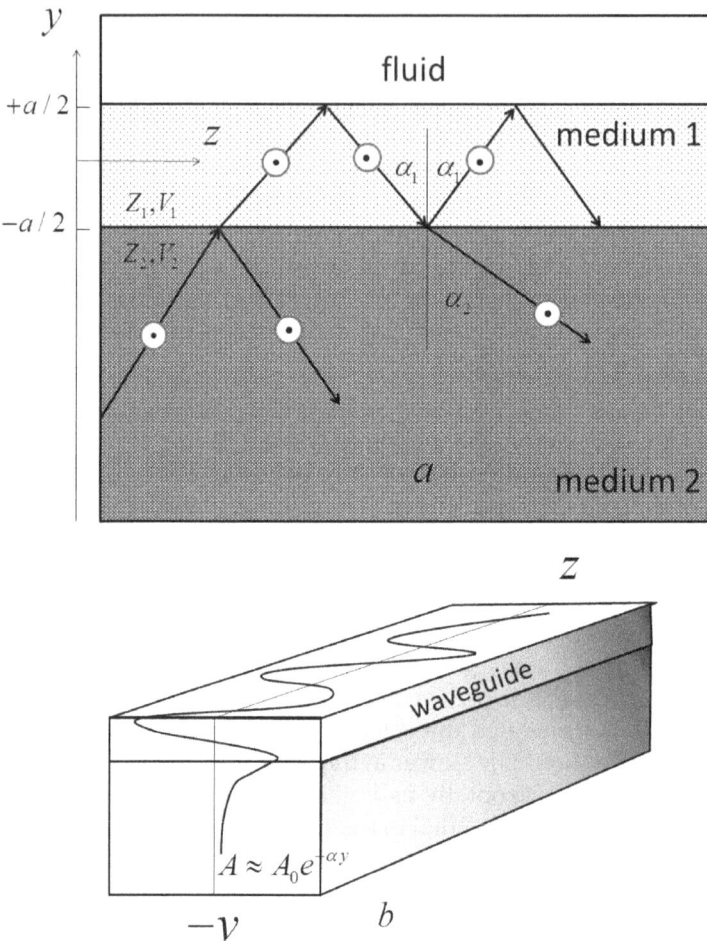

Fig. 3.3.1. (a) SH-type Love waves propagate in the layer carrying the waveguide; (b) SH-type Love waves attenuate exponentially in the depth of the waveguide.

$$T_{SH}^{SH} = \frac{2Z_1 \cos \theta_I^{SH}}{Z_1 \cos \theta_I^{SH} + Z_2 \cos \theta_T^{SH}} = \frac{v_{xT}^{(2)}}{v_{xI}^{(1)}} = \frac{B^{(2)} e^{-ik_{tSH}^{(2)} b/2}}{A e^{-ik_{tSH}^{(1)} b/2}} \qquad (3.3.4)$$

The transverse resonance condition requires that the two reflection coefficients Eq. 3.3.2 and Eq. 3.3.3 are satisfied simultaneously:

$$\frac{Z_1 \cos \theta_I^{SH} - Z_2 \cos \theta_T^{SH}}{Z_1 \cos \theta_I^{SH} + Z_2 \cos \theta_T^{SH}} = \frac{e^{ik_{tSH}^{(1)} b}}{e^{-ik_{tSH}^{(1)} b}} \qquad (3.3.5)$$

Equation 3.3.5 leads to:

$$i \tan k_{tSH}^{(1)} b = \frac{V_{SH}^{(2)} Z_{SH}^{(2)} k_{tSH}^{(2)}}{V_{SH}^{(1)} Z_{SH}^{(1)} k_{tSH}^{(1)}} \tag{3.3.6}$$

$k_{tSH}^{(2)}$ is a real number if the transmitted wave carries elastic energy away from the plate and the solution is a leaky wave. Since solutions which trap the elastic energy in the waveguide are of interest only $k_{tSH}^{(2)} = -i\alpha_{tSH}^{(2)}$ where $\alpha_{tSH}^{(2)}$ is the attenuation coefficient of the SH-wave in the half space. Therefore Eq. 3.3.6 becomes:

$$\tan k_{tSH}^{(1)} b = \frac{V_{SH}^{(2)} Z_{SH}^{(2)} k_{tSH}^{(2)}}{V_{SH}^{(1)} Z_{SH}^{(1)} k_{tSH}^{(1)}} \tag{3.3.7}$$

The solution gives two relations:

$$\left(k_{tSH}^{(1)} \right)^2 = \left(\frac{\omega}{V_{SH}^{(1)}} \right)^2 - \beta^2 \tag{3.3.8}$$

and

$$\left(\alpha_{tSH}^{(2)} \right)^2 = \beta^2 - \left(\frac{\omega}{V_{SH}^{(2)}} \right)^2 \tag{3.3.9}$$

Equation 3.3.9 shows that trapping of elastic energy can occur only if $V_{SH}^{(2)} > V_{SH}^{(1)}$. Analyzing the dispersion curves of the Love waves from Eq. 3.3.7 Auld [13] has demonstrated that the branches of the tangent function in the plot of $(\alpha_{tSH}^{(2)})^2$ as a function of $(k_{tSH}^{(1)})^2$ correspond to various Love wave modes. The attenuation coefficient $(\alpha_{tSH}^{(2)})^2 = 0$ only for the fundamental Love wave mode $n = 0$ at $\omega = 0$. All other modes are leaky waves leaking energy into the half space substrate. At higher frequency $\omega \to \infty$ the Love waves approach regular SH waves.

These results have an important impact on the theory of the seismic Love waves which propagate at very low frequency. The Earth's crust can serve as such a plate in which SH-waves propagate as in a waveguide because of total reflection. The elastic impedance of the rock layers under the crust could easily be bigger than the elastic impedance of the crust. If bulk elastic waves coming from these rock layer reach the crust at an angle bigger than the critical one Love waves can form in the crust waveguide. Because of their waveguide propagation Love waves can carry significant power at longue distances and cause devastating damage to buildings. However, such geological waveguide-like structures are rare. The probability of growth of Love-like waves with horizontal polarization called skimming waves is much higher than that of Love waves because skimming waves do not require any waveguide structure. Skimming waves will be discussed in Section 3.7.3.

Trapped Love waves propagating in the waveguide are excited only if $V_{SH}^{(2)} > V_{SH}^{(1)}$. It is interesting to analyze also what happens in the case when SV-waves propagate. In this case we have the so called generalized Lamb waves. However, as we mentioned already calculating the reflection and refraction coefficients of SV-waves is much more complicated task than the SH-wave. The dispersion relations can be found in (Auld 1973). As in the case of Love waves, the generalized Lamb waves' properties depend strongly on the ratio $V_{SV}^{(1)}/V_{SV}^{(2)}$. In the limit case when $V_{SV}^{(1)} \gg V_{SV}^{(2)}$ there is only one generalized Lamb wave solution, which reduces to a Rayleigh wave on the surface between the plate and the substrate when $\beta b \to 0$ and exists only if $\omega/\beta < V_{SV}^{(2)}$. If $V_{SV}^{(1)} \ll V_{SV}^{(2)}$ there is an infinite number of solutions which can be grouped into two families of modes called the M_1 and M_2 series. The fundamental modes of these series—(M_{11}, M_{21})—are the most interesting. For a plate thickness close to zero or for very low frequencies ($\beta b \to 0$) the mode M_{11} becomes a Rayleigh type of wave, while the higher orders of M_1 and all M_2 modes become leaky waves. As the plate thickness increases, the first trapped mode is the M_{21} called Sezawa wave. It is interesting that Sezawa waves have not been considered in seismology, despite that fact that there the SV-cousins of the SH Love waves. Often Sezawa waves are confused with Rayleigh waves, but there is a difference. Rayleigh waves are couples SV- and P-waves, while Sezawa waves are pure SV-wave without any P-wave coupled to them. For thick plate or high frequency ($\beta b \to \infty$) the M_{11} mode becomes a Rayleigh wave on the upper boundary of the plate between the plate and the fluid and all other higher modes including the M_{21} degenerate into SV-waves in the plate. For certain material impedances when $V_{SV}^{(1)} \approx V_{SV}^{(2)}$ become a bound surface wave on the plate-substrate interface. Waves of such type may exist not only when $V_{SV}^{(1)} \approx V_{SV}^{(2)}$ but also $V_{SV}^{(1)} > V_{SV}^{(2)}$.

3.4 Stoneley waves

As discussed in the previous section for specific material parameters when $\beta b \to 0$ the M_{21} mode becomes a surface wave bound to the interface between the plate and the half space substrate. This case corresponds to material combinations where $\hat{V}_S \approx V_S$. This interface wave is called Stonley wave. The Stonley wave can propagate only if its velocity is bigger than the velocity of Rayleigh wave's velocity and smaller than the velocity of the S-waves in the denser medium.

$$V_R^{(2)} < V_{SV}^1 < V_{SV}^2 \tag{3.4.1}$$

As we discussed it is difficult to calculate V_{SV}^1 without numerical computation. However, it is certain that its value has to be between $V_R^{(2)}$ and V_{SV}^2 when $\rho^{(2)} > \rho^{(1)}$. The region of existence of Stoneley wave is shown in Fig. 3.4.1.

The existence of Stonley waves is restricted by the elastic properties of medium 1 and medium 2. In homogeneous materials Stonley waves arise rarely. In heterogeneous materials such as the Earth's crust composed by a great number of interfacing domains with different elastic properties and elastic wave velocities Stonley waves can be easily created. Stonley waves propagate on the domains' boundaries with both P-components and S-components. They are exponentially evanescent on the two sides of the boundary as shown in Fig. 3.4.2 for W/Al interface (Farnell and Adler 1970).

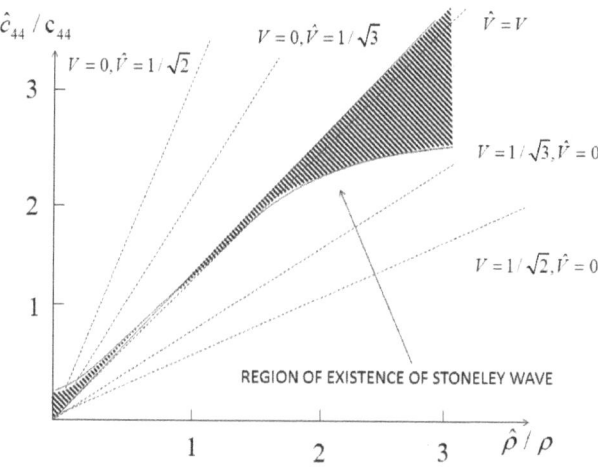

Fig. 3.4.1. Region of existence of Stoneley wave (Farnell and Adler 1972).

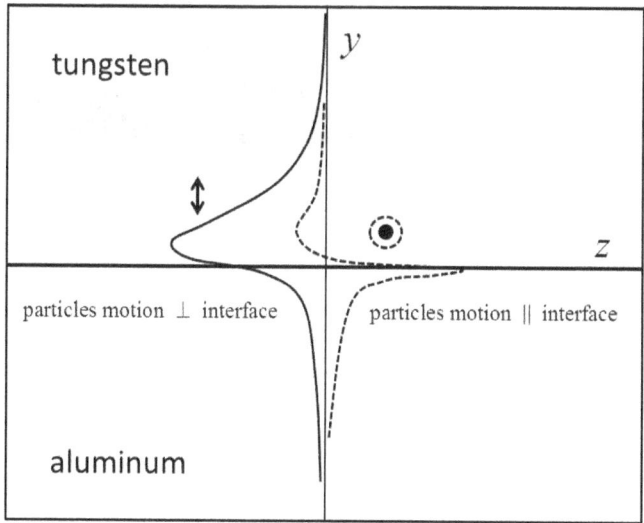

Fig. 3.4.2. Stoneley waves at W/Al interface.

Stonley waves can also be generated at liquid-solid interface. These waves are called Scholte waves or borehole waves because they arise often during oil and gas drilling operations.

3.5 Reflection and transmission of SAW

Such reflection of the surface elastic wave can be produced by a vertical step on the surface such as the edge of the substrate in Fig. 3.5.1 (Farnell 1978). It can be calculated that 41% of the incident surface elastic wave energy will be transmitted around the corner and will propagate down the vertical surface, 13% will be reflected back as a surface elastic wave propagating away from the edge, and 46% of the incident wave will be converted into bulk elastic modes propagating down the substrate away from the edge. Figure 3.5.2 shows the percentage of incident surface elastic wave energy reflected and transmitted as Rayleigh waves (dashed curve) and converted to bulk waves (solid curve) as a function of the ratio step height vs. wavelength (Farnell 1978).

With these results if we go back to Fig. 3.5.1 we can conclude that if such steps made of concrete, metal, plastic or other appropriate material with appropriate values of h are built around a city in the case of an earthquake cause by Rayleigh surface seismic waves almost the entire amount of seismic energy carried by the waves could be converted, reflected, and attenuated. In the next section of this book we will continue exploring this idea. We have to answer the most important question: how to design such steps so they can be efficient in a wide frequency band (a wide range of wavelengths).

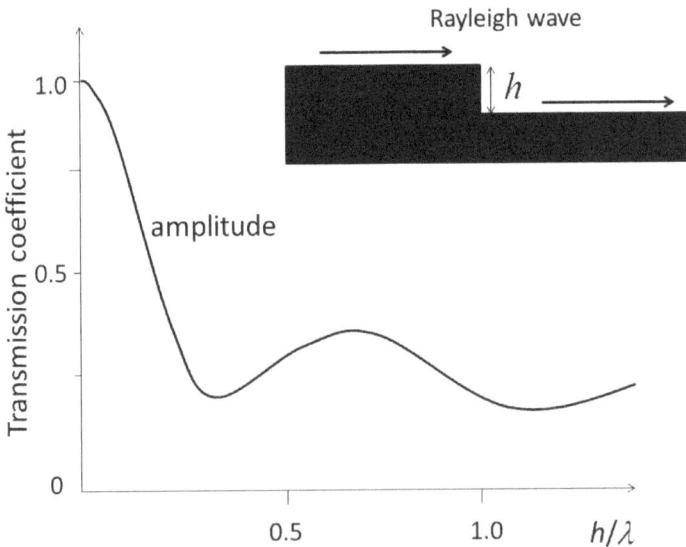

Fig. 3.5.1. Amplitude of the transmitted Rayleigh wave normally incident on a step (Farnell 1978).

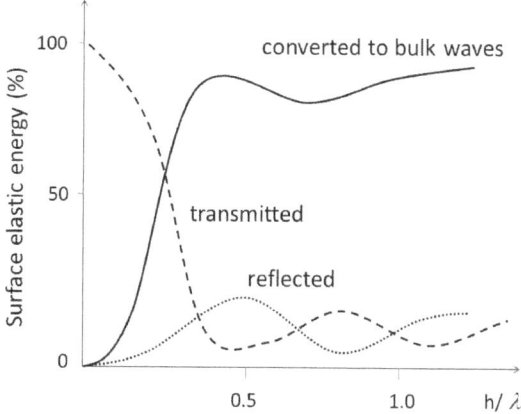

Fig. 3.5.2. Reflected, transmitted, and converted Rayleigh waves from a step as a function of step's height vs. wavelength (Farnell 1978).

3.6 Attenuation and scattering of SAW

Attenuation and scattering of surface elastic waves happens usually in heterogeneous media where seismic waves propagate. Rocks are strongly nonlinear because of their attenuation and scattering properties. These properties make them to be highly dispersive. The resulting nonlinear phenomena change dramatically the propagation of surface elastic waves.

3.6.1 Attenuation of SAW

Bulk and surface elastic waves lose energy during their propagation. The energy loss depends on the composition, elastic properties, homogeneity, and temperature of the medium of propagation on one hand, and on the polarization and frequency of the elastic wave on the other hand. The propagation energy loss in the medium of propagation is due to a combination of three factors (Slobodnik 1978): 1) Interaction of thermally generated elastic waves; 2) Scattering by defects, inclusions, and other inhomogeneity factors; 3) Energy loss to fluids adjacent to the boundary surface. The first factor is temperature dependent, the second one is temperature independent, and the third one is pressure dependent.

The impact of these three factors and their combinations on the attenuation of bulk and surface waves is different in different frequency ranges. While the first factor has a significant impact on surface elastics waves in the microwave frequency range used in physical acoustics because of interactions with thermal phonons, this factor can be neglected for the surface seismic waves on the Earth's surface except for bulk waves propagating deep in the mantle and core where temperatures are high. The highly inhomogeneous Earth's structure has a significant impact on the propagation loss of bulk and surface seismic waves, and little impact on elastic waves in physical acoustics which usually propagate in highly homogeneous materials. The third factor affects high-frequency surface elastic waves in physical acoustics. It has small impact on Love and Rayleigh surface seismic waves, but a significant one on Stoneley surface elastic waves and bulk seismic waves propagating around the Earth's liquid core.

Surface elastic wave attenuation problem can be analyzed using the perturbation theory (Auld 1973). Consider attenuation of Rayleigh surface waves from surface roughness shown in Fig. 3.6.1.1. The surface roughness is characterized by a roughness function $f_r(z)$ and a roughness parameter ε. Also we consider that $y_r(z) \ll \lambda_{R'}$ where λ_R is Rayleigh wavelength, and the slope of the perturbed surface is small, i.e., $df_r(z)/dz \ll 1$. Therefore we have:

$$\hat{n}(z) = \hat{y} n_y(z) + \hat{z} n_z(z) \approx \hat{y} - \hat{z}\varepsilon \frac{df(z)_r}{dz} \qquad (3.6.1.1)$$

Since the surface is stress free the boundary condition is:

$$\vec{T}(y,z) \cdot \hat{n}(z) = \vec{T}(y,z) \cdot \hat{y} - \varepsilon \frac{df_r(z)}{dz} \vec{T}(y,z) \cdot \hat{z} = 0, \; y = y_r(z) \qquad (3.6.1.2)$$

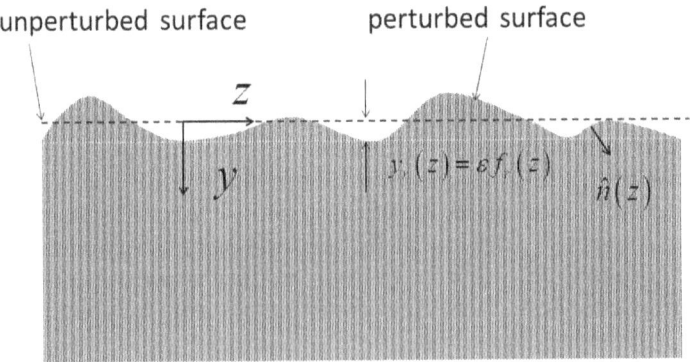

Fig. 3.6.1.1. Rough surface characterized by a roughness function $f(z)$ and perturbation parameter.

By expanding $\vec{T}(y, z)$ in powers of ε we get:

$$\vec{T}(y,z) = \vec{T}^{(0)}(y,z) + \varepsilon \vec{T}^{(1)}(y,z) + \ldots$$

The boundary condition Eq. 3.6.1.2 can be written as:

$$\vec{T}(y,z) \cdot \hat{n}(z) = \vec{T}^{(0)}(y,z) \cdot \hat{y} + \varepsilon \left(\vec{T}^{(1)}(y,z) - \frac{df_r(z)}{dz} \vec{T}^{(0)}(y,z) \cdot \hat{z} \right) + \ldots = 0$$

$$(3.6.1.3)$$

The zero order term in Eq. 6.3.2.3 is zero because it corresponds to the unperturbed boundary condition and the first order term is:

$$\left(\vec{T}^{(1)}(y,z) \cdot \hat{y} \right)_{y=0} = -f_r(z) \left(\frac{\partial \vec{T}^{(0)}(y,z) \cdot \hat{y}}{\partial y} \right)$$

$$(3.6.1.4)$$

The average power density loss per unit surface P_{loss} of the Rayleigh wave can be calculated using Eq. 3.6.1.4. If P_R is the average Rayleigh power density per unit width normal the z-axis the attenuation is:

$$\alpha_R = (20 \log e) \frac{P_{loss}}{P_R} \text{ dB/m}$$

$$(3.6.1.5)$$

By expanding the roughening function $f_r(z)$ in Fourier series the attenuation coefficient α_R can be obtained.

$$f_r(z) = \sum_{m=-\infty}^{\infty} f_{rm} e^{i 2 \pi m z / \lambda_r}$$

$$(3.6.1.6)$$

In Eq. 3.6.1.6 λ_r is the roughness period of the surface. The attenuation coefficient α_R is a sum of the attenuation coefficients of the individual Fourier components:

$$\alpha_R = \sum_{m=-\infty}^{\infty} \alpha_{Rm}$$

$$(3.6.1.7)$$

For an isotropic surface the m-th Fourier component gives the attenuation expression of Brekhovskikh:

$$\alpha_{Rm} = \frac{4\pi^3}{\lambda_s} \frac{(s-1)\sqrt{s(s-1)} \left[\sqrt{q - \eta^2} |L|^2 + \sqrt{1 - \eta^2} |T|^2 \right]}{4s^2 - 3qs - 3s + 2q - 2(2s-1)\sqrt{(s-q)(s-1)}} \text{ dB/m} \quad (3.6.1.8)$$

where with λ_p, λ_S, and λ_R being the wavelengths of the P-wave, S-wave, and Rayleigh wave

$$s = \left(\lambda_S / \lambda_R\right)^2$$

$$q = \left(\lambda_S / \lambda_P\right)^2$$

$$\eta = \left(\lambda_S / \lambda_R\right) + m\left(\lambda_S / \lambda_r\right)$$

$$L = 2\frac{f_{rm}}{\lambda_S \Delta}\left\{\frac{\left(1-2\eta^2\right)\left(2s-1\right)}{4s\sqrt{s-1}} + 2i\eta\sqrt{1-\eta^2}\left[\frac{\left(1+4sq-4s\right)}{2\sqrt{s}} - \frac{2m\lambda_S}{\lambda_r}\left(1-q\right)\right]\right\}$$

$$T = 2\frac{f_{rm}}{\lambda_S \Delta}\left\{\frac{\eta\left(2s-1\right)\sqrt{q-\eta^2}}{2s\sqrt{s-1}} + i\left(2\eta^2-1\right)\left[\frac{\left(1+4sq-4s\right)}{2\sqrt{s}} - \frac{2m\lambda_S}{\lambda_r}\left(1-q\right)\right]\right\} \quad (3.6.1.9)$$

$$\Delta = -\left(2\eta^2-1\right)^2 - 4\eta^2\sqrt{\left(q-\eta^2\right)\left(1-\eta^2\right)}$$

Figures 3.6.1.2 and 3.6.1.3 show the attenuation coefficient of the Rayleigh wave propagating on saw tooth surfaces with angles of 10° and 25° (Sabine 1970; Rischbieter et al. 1965).

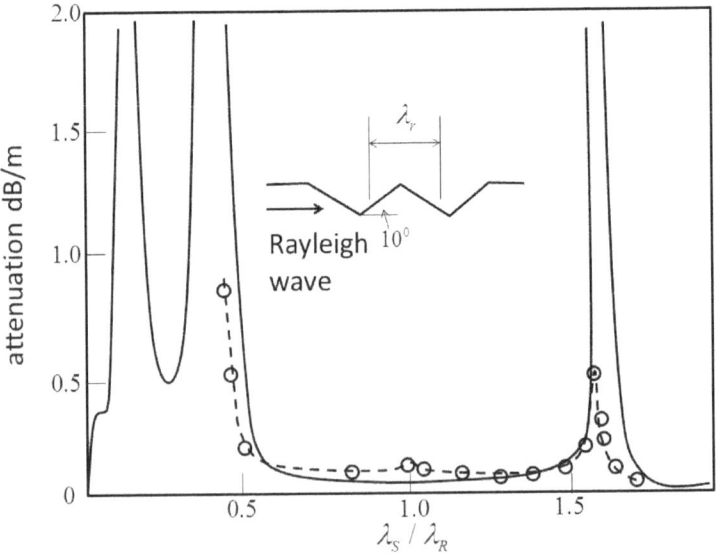

Fig. 3.6.1.2. Rayleigh wave attenuation due to saw tooth roughness function.

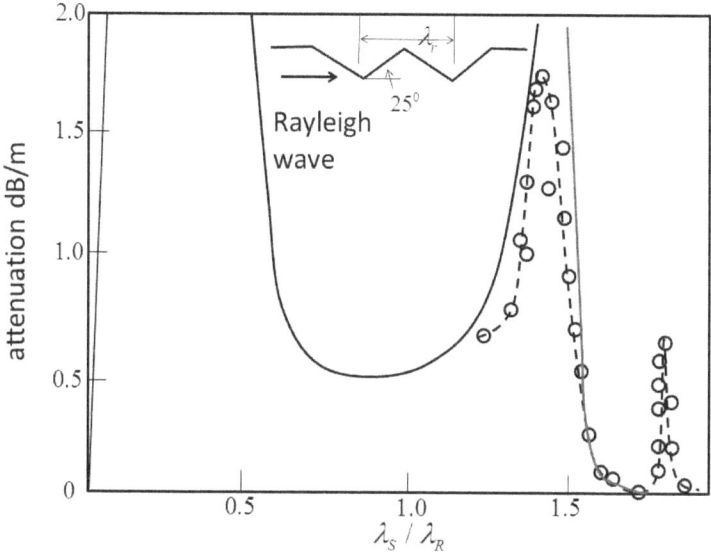

Fig. 3.6.1.3. Rayleigh wave attenuation due to saw tooth roughness function.

3.6.2 Excitation of Raleigh waves by body waves

Scattering is a physical process that causes radiation to deviate from a straight trajectory (Slobodnik 1978). If there are microscopic irregularities in the surface an incident optical beam will get diffuse instead of specular reflection. The same happens for light passing through a transparent medium. If there are non-uniformities like particles or bubbles some of the light will deviate from its original trajectory. There are two types of scattering: elastic and inelastic. During the process of elastic scattering there is no loss or gain of energy by the optical wave. During the process of inelastic scattering there is a change in the energy of the wave. If the light is substantially extinguished by the interaction (losing a significant proportion of its energy), the process is known as absorption. Reflection of light is an elastic process.

A single scattering happens when there is only one scattering center. Often many randomly located scattering centers are present in the medium resulting in a multiple scattering process. Since the distribution of the scattering centers is unknown it is difficult to solve the 'inverse scattering problem'—by measuring the intensity of the scattered light to find either the intensity of the incident light or the distribution of the scattering centers.

Light scattering and absorption are the two major physical processes that contribute to the visible appearance of physical objects. The spectral distribution of absorption determines the color of a surface, while the amount of scattering determines whether the surface is mirror-like or not (Fig. 3.6.2.1). The wavelength dependence of scattering is determined by the ratio $x = 2\pi r/\lambda$ of r—the characteristic length (radius of the particle) and λ—the optical wave wavelength. If $x \ll 1$ (the size of the particles is much smaller than the wavelength) is called Rayleigh scattering. If $x \approx 1$ the scattering is called Mie scattering. If $x \gg 1$ (the size of the particles is much larger than the wavelength) the shape of the scattering object becomes much more significant. Short wavelength light is scattered by the atmosphere much more than long wavelength light.

Elastic waves similarly to electromagnetic waves are scattered by objects, however the scattering is usually inelastic because there are always transmitted waves except for some special cases of critical angles when there is no transmission. Not only transmitted waves cause the non-linearity of the process, but also the reflection itself is not linear because additionally to the reflected wave of same polarization another type of wave appears. If the boundary surface is rough and the size of the rough areas is of the order of the elastic wavelength or larger scattering will complicate the process even more. Bulk and surface elastic wave scattering from the random rough boundaries between various domains in the crust occur much more often than reflection and transmission of plane waves from smooth surface that have been discussed in the previous sections. Most of the surfaces and boundaries between domains in the Earth's crust are neither plane nor smooth; they are rough and curved. In this case a plane elastic wave incident to that surface will be scattered in various directions as elastic waves with modified polarizations and wave fronts.

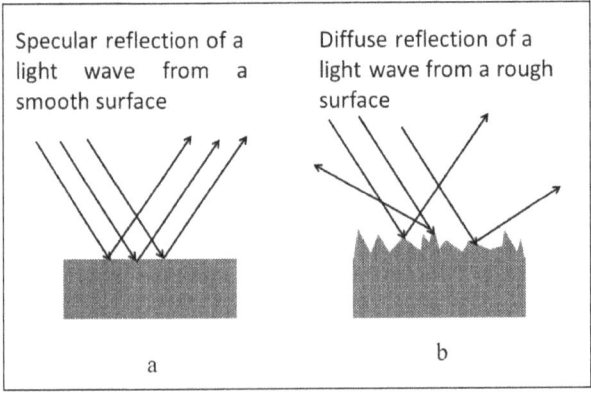

Fig. 3.6.2.1. Specular reflection and diffuse reflection of optical waves.

The problem of scattering of acoustic waves from random surfaces or solids dispersed in the medium of propagation has been object of active research during the past couple of decades both experimentally and theoretically. The theory of elastic waves scattering is extremely complex and to date there is no coherent solution to this problem. The problem of reflection and transmission of elastic waves in anisotropic solids is also very complex. Only reflection and transmission in isotropic media has been fully analyzed and described in literature. Various methods to analyze the scattering problem and various approximations have been proposed. A detailed review of the research done in this area has been done elsewhere (Ogilvy 1987).

Theoretical results (Gilbert and Knopoff 1960) confirmed experimentally (Hudson et al. 1973) show that Rayleigh surface waves incident to a triangular grove is scattered into bulk longitudinal waves and vice versa (Figs. 3.6.2.2 and 3.6.2.3). Figure 3.6.3.3 shows a set of triangular groves converting an incident Rayleigh surface wave into longitudinal bulk elastic waves. The amplitudes of the longitudinal waves depend on the grove angle. The amplitude of the scattered waves varies also with the scattering angle.

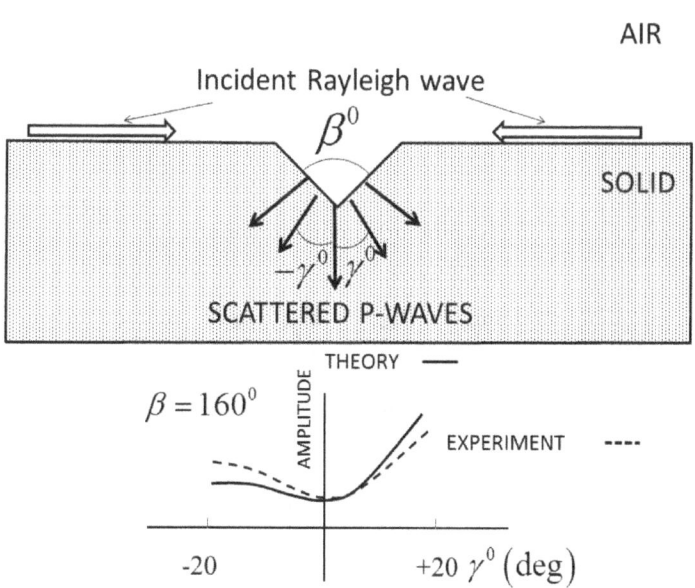

Fig. 3.6.2.2. A triangular grove converts Rayleigh surface waves into longitudinal bulk elastic waves.

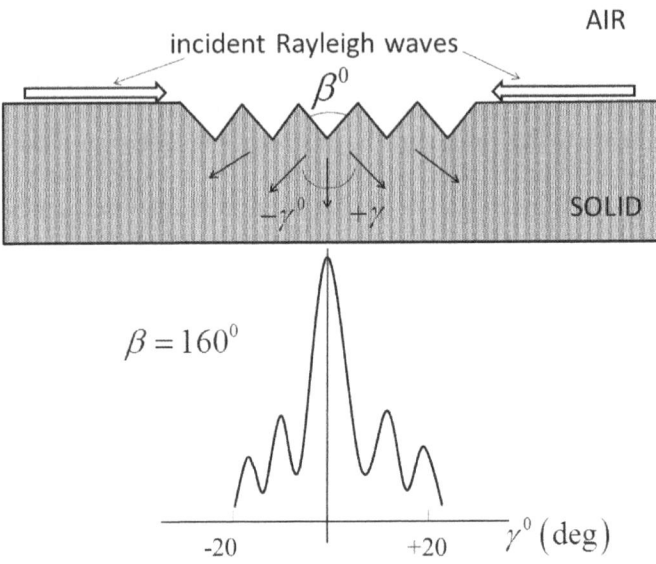

Fig. 3.6.2.3. A set of triangular groves converts Rayleigh surface waves into longitudinal bulk waves.

Rayleigh surface waves can be excited by plane bulk wave scattering at a surface with periodic mass loading as shown in Fig. 3.6.2.4.

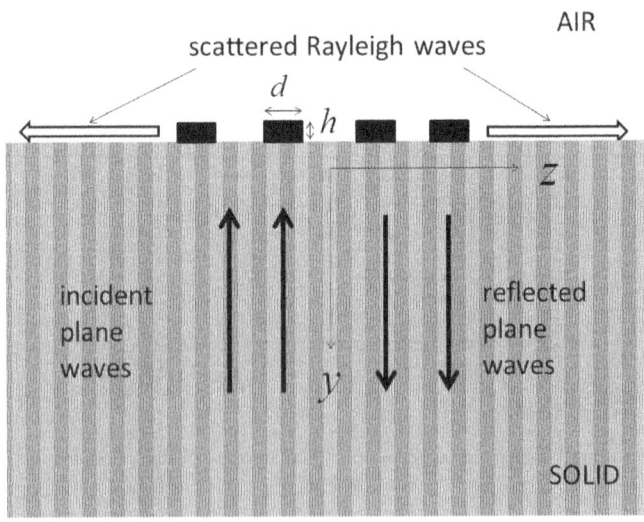

Fig. 3.6.2.4. A set of triangular groves converts Rayleigh surface waves into longitudinal bulk waves.

The boundary conditions are defined by:

$$-\left(\vec{T}^{(1)}(z)\cdot\vec{y}\right)_{y=0} = \Delta\vec{Z}\cdot\left(\vec{V}^{(0)}\right)_{y=0}$$

If only mass loading is considered

$$\Delta Z = -i\omega h\rho'(z)$$

The scattered Rayleigh waves are given by the relations (Auld 1973):

$$\left(\frac{d}{dz}+i\beta_R\right)A_{R+}(z) = \frac{1}{4P_R}\left[-V_{R+}^*\cdot\vec{T}^{(1)}(z)\cdot\vec{y}\right]_{y=0}$$

$$\left(\frac{d}{dz}-i\beta_R\right)A_{R-}(z) = -\frac{1}{4P_R}\left[-V_{R0-}^*\cdot\vec{T}^{(1)}(z)\cdot\vec{y}\right]_{y=0}$$

(3.6.2.1)

Evaluating the wave amplitudes details from Eq. 3.6.1.10 can be found in (Auld 1973). Basically the power radiated from the bulk P-waves into Raleigh waves is:

$$P_+ = |A_{R+}|^2 P_P$$

$$P_- = |A_{R-}|^2 P_P$$

(3.6.2.2)

3.7 Nonlinear surface elastic waves in nondispersive media

Surface elastic waves that propagate on an isotropic homogeneous surface are nondispersive. If the medium is nonlinear the absence of dispersion allows the surface wave to create a great number of harmonics. Higher Rayleigh wave harmonics have been observed experimentally by many authors (Mayer 1995). Using the method of multiple scales Lardner (Lardner 1984) has derived an evolution equation of waveform of a Rayleigh wave caused by second-order nonlinear terms. The method has been developed further by Parker (Parker 1988).

3.7.1 Equation of evolution

The Hamiltonian of a linear system is the sum of the potential and kinetic energy. The same is in the case of a nonlinear elastic system, however now the potential energy includes higher orders of the displacement as we discussed in Chapter 2 (Eq. 2.5.1.9). Equation 3.1.8 is a solution of the equation of propagation with boundary conditions corresponding to elastic wave propagation on the (x_1, x_2)—free surface and vanishing in the direction of $-x_3$. If the propagation is in the direction of x_1, following the analysis of Hamilton

(Hamilton et al. 1999) the kinetic and potential energies can be presented in terms of harmonic amplitudes $a_n(t)$ chosen to be generalized coordinates, in which the corresponding momenta according to Eq. 2.5.2.10 are $p_i = \partial E_k / \partial \dot{a}_i$. Equation 3.1.8 can be presented in the form.

$$u_i(x_1, x_3, t) = \sum_n a_n(t) u_{ni}(z) e^{ik_x x_1} \tag{3.7.1.1}$$

where

$$u_{ni}(x_3) = \sum_{s=1}^{3} C_s \alpha_i^{(s)} e^{ink\xi_s} \text{ and } u_{ni} = u_{(-n)i}^* \tag{3.7.1.2}$$

The Hamilton equations Eq. 2.5.2.21 become:

$$\dot{a}_n = \frac{\partial H}{\partial p_n} \text{ and } p_n = -\frac{\partial H}{\partial a_n} \tag{3.7.1.3}$$

The kinetic energy density (per unit area) is:

$$E_k = \frac{\rho}{2\lambda} \int_{x_1}^{x_1+\lambda} \int_{-\infty}^{0} \dot{u}_i \dot{u}_i dx_1 dx_3 \tag{3.7.1.4}$$

In Eq. 3.7.1.4 $\lambda = 2\pi k$ is the wavelength corresponding to the fundamental mode in the Fourier expansion of u_i. With Eq. 3.7.1.1, Eq. 3.7.1.4 becomes (Hamilton et al. 1999):

$$E_k = \frac{\rho}{2} \sum_n \dot{a}_n \dot{a}_{-n} \int_{-\infty}^{0} |u_{ni}|^2 dx_3 = \frac{\rho}{2k} \sum_n \frac{\dot{a}_n \dot{a}_{-n}}{|n|} \tag{3.7.1.5}$$

The quadratic part of the potential energy is:

$$E_P^{(q)} = \frac{1}{2\lambda} \int_{x_1}^{x_1+\lambda} \int_{-\infty}^{0} u_i \frac{\partial T_{ij}}{\partial x_j} dx_1 dx_3 \tag{3.7.1.6}$$

With Eq. 2.5.2.11 and Eq. 3.7.1.1 the potential energy Eq. 3.7.1.6 becomes:

$$E_P^{(q)} = \frac{1}{2} \rho V_R^2 k \sum_n |n| a_n a_{-n} \tag{3.7.1.7}$$

Following the same path to determine the quadratic terms of the potential energy the cubic term Eq. 2.5.1.9 can be expressed as:

$$E_P^{(c)} = \frac{c_{ijklmn}}{6k} \int_{x_1}^{x_1+\lambda} \int_{-\infty}^{0} \frac{\partial u_i}{\partial x_j} \frac{\partial u_k}{\partial x_l} \frac{\partial u_m}{\partial x_n} dx_1 dx_3 = \sum_{n_1 n_2 n_3} w_{n_1 n_2 n_3} a_{n_1} a_{n_2} a_{n_3} \tag{3.7.1.8}$$

We consider propagation of surface elastic waves on the free boundary of a solid filling the space $x_3 < 0$ (Mayer 2008). The surface wave displacement u_i will depend on the coordinates x_1 and x_2 and time t. Piola-Kirchhoff stress tensor T_{ij} describes conveniently the nonlinear phenomena:

$$T_{ij} = \sum_{k,l=1}^{3} c_{ijkl} u_{kl} + \frac{1}{2} \sum_{k,l,m,n=1}^{3} c_{ijklmn} u_{kl} u_{mn} + \ldots \qquad (3.7.1.9)$$

The equation of motion is:

$$\rho \ddot{u}_\alpha = \sum_{\beta=1}^{3} \frac{\partial T_{\alpha\beta}}{\partial x_\beta} \qquad (3.7.1.10)$$

At the stress-free surface $x_3 = 0$ the boundary condition is $T_{\alpha 3} = 0$ and because we deal with a surface wave the displacement will become 0 if $x_3 \to -\infty$.

In the linear limit the solution of Eq. 3.7.1.10 will be (Mayer 2008):

$$u_{\alpha,1}(x_1, x_3, t) = e^{ik(x_1 - V_R t)} \Psi_\alpha(x_3 | k) B(k) + c.c. \qquad (3.7.1.11)$$

In Eq. 3.7.1.11 $\Psi_\alpha(x_3|k)$ is a depth profile function and V_R is the Rayleigh wave phase velocity. $\Psi_\alpha(x_3|k)$ is normalized in such a way that $B(k)$ is the Fourier transform of gradient displacement components $u_{\alpha\beta}$ at the free boundary $x_3 = 0$. The displacement can be presented in the form (Parker 1988; Parker et al. 1992; Mayer 1995):

$$\bar{u}(x_1, t) = \varepsilon \bar{u}\left(\theta, x_3, x_1^{(1)}\right) + \varepsilon^2 \bar{u}\left(\theta, x_3, x_1^{(1)}\right) + O\left(\varepsilon^3\right) \qquad (3.7.1.12)$$

In Eq. 3.7.1.12 ε is the weak-nonlinearity asymptotic-perturbation expansion parameter, $\theta = x_1 - V_R t$ and $x_1^{(1)} = \varepsilon x_1$ is the stretched coordinate. A stretched time or a spatial coordinate is introduced that represents the scale on which a waveform evolution takes place $\tau = \varepsilon t = \varepsilon x_1 / V_R$.

The compatibility condition for the solvability of the system, obtained by projecting on a linear surface wave solution, yields the evolution equation for the strain amplitudes B as a function of the wave number k in the following form:

$$i\frac{\partial B(k)}{\partial \tau} = \frac{V_R k}{2\pi} \left\{ \int_0^q F(q/k) B(q) B(k-q) dk + \right.$$

$$\left. 2\int_k^\infty (k/q) F^*(k/q) B(q) B^*(q-k) dk \right\} \qquad (3.7.1.13)$$

The dimensionless function F in Eq. 3.7.1.5 is an overlap integral of linear surface waves involving c_{ijklmn} and the depth profile function $\Psi_\alpha(x_3|k)$.

In isotropic homogeneous half-space the strain amplitude B in the evolution Eq. 3.7.1.13 is the Fourier transform of $u_{3,1}$ at the free surface, i.e., the local surface slope, then the dimensionless function F is imaginary. In the general case, however, F has both a non-vanishing real and imaginary part.

By inserting Eq. 3.3.4 into the equation of motion Eq. 3.1.11 with the surface wave boundary conditions and retaining terms only of first order in ε we get a solution of the linearized equation of motion in the form:

$$\bar{u}\left(\theta, x_3, x_1^{(1)}\right) = \sum_k \alpha\left(k, x_1^{(1)}\right) \varpi\left(x_3 |k\right) e^{ikt} \qquad (3.7.1.14)$$

$\varpi(x_3|k)$ is a function of the displacement field of the Rayleigh wave in the homogeneous half space.

With Eq. 3.3.6 the equation of motion leads to the equation of evolution for the amplitude $\alpha(k, x_1^{(1)})$

$$i \frac{\partial \alpha\left(k, x_1^{(1)}\right)}{\partial x_1^{(1)}} = \sum_{k'} K\left(k, k'\right) \alpha\left(k, x_1^{(1)}\right) \alpha\left(k - k', x_1^{(1)}\right) \qquad (3.7.1.15)$$

$K(k, k')$ is given by:

$$K\left(k, k'\right) = \frac{V_R |k|}{2 s_R\left(k\right) s_{jJmMnN}} \int_{-\infty}^{0} \left[\Theta_J\left(k\right) \varpi_J\left(x_3 |k\right)\right]^* \left[\Theta_M\left(k'\right) \varpi_J\left(x_3 |k\right)\right] \times$$

$$\left[\Theta_N\left(k - k'\right) \varpi_J\left(x_3 |k - k'\right)\right] dx_3 = \tilde{K}\left(k'|k\right) k'\left(k - k'\right) \qquad (3.7.1.16)$$

In Eq. 3.7.1.16 we have:

$$S_{aA} = \delta_{aA} C_{AB}$$

$$S_{aBcD} = \delta_{aA} \delta_{cD} C_{ABCD} + \delta_{ac} C_{BD}$$

$$S_{aBcDeF} = \delta_{aA} \delta_{cC} \delta_{eE} C_{ABCDEF} + \delta_{aA} \delta_{ce} C_{ABDF} + \delta_{cC} \delta_{ae} C_{CDBF} + \delta_{eE} \delta_{ac} C_{BDEF}$$

Also we have:

$$\Theta_J\left(k\right) = \delta_{J1} ik + \delta_{J3} \frac{\partial}{\partial x_3} \text{ and } s_R\left(k\right) = 2V_R |k| s_{J1mM} \text{ Im} \int_{-\infty}^{0} \varpi_J^*\left(x_3 |k\right) \Theta_M\left(k\right) \varpi_m\left(x_3 |k\right) dx_3$$

Surface elastic waves propagating on the half-space surface of an elastic, homogeneous, and nondispersive medium are nondispersive. If the propagation is nonlinear because either the medium is nonlinear or the wave has finite amplitude the initially sinusoidal surface wave creates higher harmonics which grow without being inhibited by dispersion.

The theory of nonlinear acoustic waves in isotropic and anisotropic media has been recently developed by (Parker 1988; Zabolotskaya 1992; Hamilton et al. 1999; Gusev et al. 1998) using Hamiltonian formalism.

The Hamiltonian is

$$H = E_k + E_p + W_{nonlinear} \qquad (3.7.1.17)$$

The displacement expressed in terms of harmonic functions $q_n(t)$ chosen to be generalized coordinates is given by:

$$u_i(x_1, x_3, t) = \sum_n q_n(t) u_{ni}(x_3) e^{inkx_1} \qquad (3.7.1.18)$$

Also with the definition of $q_n(t)$ we have $p_n = \dfrac{\partial E_k}{\partial \dot{q}_n}$ and the Hamilton's canonical equations become Eq. 2.5.2.21.

The evolution equation in terms of particle displacement velocity rather than displacement as function of propagation distance is given by (Mayer 2008):

$$v_n = \frac{n^2 \omega}{2\rho V^3} \left(\sum_{m=1}^{n-1} S_{m,n-m} v_m v_{n-m} - 2 \sum_{m=n+1}^{\infty} S^*_{n,m-n} v_m v^*_{m-n} \right) \qquad (3.7.1.19)$$

The nonlinearity matrix in Eq. 3.7.1.11 is given by:

$$S_{n_1 n_2} = \sum \frac{F_{S_1 S_2 S_3}}{n_1 \xi_{S_1} + n_2 \xi_{S_2} - (n_1 + n_2) \xi^*_{S_3}} \qquad (3.7.1.20)$$

In Eq. 3.7.1.20 $F_{S_1 S_2 S_3}$ is a coordinate transformation matrix for second and third-order stiffness constants c_{ijkl} and c_{ijklmn}.

Equation 3.7.1.19 is useful only in the case when the waveform is known in time. Usually it is more important to know the evolution of the waveform as a function of the propagation distance. In this case $v_n = v_n(x)$ for isotropic media Eq. 3.7.1.19 is replaced by:

$$\frac{dv_n}{dx} = \frac{n^2 \omega}{2\rho V^3} \left(\sum_{m=1}^{n-1} S_{m,n-m} v_m v_{n-m} - 2 \sum_{m=n+1}^{\infty} S^*_{n,m-n} v_m v^*_{m-n} \right) \qquad (3.7.1.21)$$

3.7.2 Wave-wave interactions in nondispersive medium

Two elastic waves can interact with each other and create a third elastic wave because of the second-order nonlinearity (Mayer 2008). At least one of the three waves involved has to be a Rayleigh wave. There could be various processes of interactions such as creation of second harmonics of Rayleigh waves or parametric mixing of two Rayleigh waves. Two Rayleigh waves with wavenumbers k_1 and k_2 can generate new Rayleigh waves with wavenumbers $k_1 \pm k_2$ in a three wave process. The amplitudes of the generated waves grow linearly with distance according to:

$$\alpha\left(k_1 \pm k_2\right) = -i2K\left(k_1 \pm k_2, k_1\right)x_1^{(1)}\alpha\left(k_1\right)\alpha\left(\pm k_2\right) \qquad (3.7.2.1)$$

Another process that can occur is a parametric amplification of a weak sub-harmonic wave by a pump wave of large amplitude (Lardner 1984; Krylov 1993). This process is governed by the equation:

$$i\frac{\partial\alpha\left(k, x_1^{(1)}\right)}{\partial x_1^{(1)}} = -iK\left(k, 2k\right)\alpha\left(2k\right)\alpha\left(-k\right) - \Gamma k^2\alpha\left(k\right) \qquad (3.7.2.2)$$

The amplitude of the pump wave has to be big enough to satisfy the condition:

$$\left|K\left(k, 2k\right)\alpha\left(2k\right)\alpha\left(-k\right)\right| > \Gamma k^2$$

It has been demonstrated (Krylov 1993) that for small attenuation the amplitude of the weak sub-harmonic wave cannot be amplified more than 2 times. If, however, the third and fourth harmonic waves are damped simultaneously, the maximum amplitude of initial Rayleigh wave can be amplified via parametric amplification up to 10 times.

Nonlinear interactions between surface and bulk waves have been studied both experimentally and theoretically. The results can of mixing nonlinear waves be summarized as follow:

  i)   a surface wave and a bulk wave generate a surface wave
 ii)   two surface waves generate a bulk wave
iii)   a surface wave and a bulk wave generate a bulk wave
iv)   two bulk waves generate a surface wave

It is interesting to discuss how nonlinearity and dispersion affect the propagation of bulk and surface seismic waves.

During distant earthquakes often two distinct stages have been observed—the first characterized by a preliminary weak motion followed by the second main shock characterized by a much stronger tremor.

First Rayleigh (Rayleigh 1885) suggested that surface seismic waves play an important role in earthquakes. Later Oldham (Oldham 1900) recorded two phases in the preliminary weak motions that he identified by their travel times as the direct bulk P- and S-waves traveling at different but almost constant velocities. The main shock Oldham attributed to Rayleigh surface elastic wave traveling also at almost constant speed on the Earth's surface but much slower than the bulk P- and S-waves. Lamb (Lamb 1904) confirmed Oldham's observations on the surface seismic waves during the main shock, but couldn't explain why in some cases the ground was moving vertically which shows a typical Rayleigh wave and in other it moved sidewinding in the horizontal ground plane. The sidewinding surface elastic waves were later were explained by Love (Love 1911) and called Love waves. As we have discussed already, instead of considering homogeneous half-space Love considered a layer on the top of the ground half space as a boundary between the ground half space and the air. Under specific conditions this layer forms an elastic wave guide. The waves that propagate in the waveguide are polarized in the horizontal plane perpendicularly to the direction of propagation. It was found out that Rayleigh waves can also propagate in a layer structure and that both Love and Rayleigh waves in a layered heterogeneous half space are dispersive waves, i.e., their velocities of propagation depend on the wave's frequency. In a homogeneous half space surface waves are nondispersive. In case of finite-amplitude propagation resonating harmonic waves can be generated and a parametric amplification and waveform evolution are observed. In a dispersive half space different phenomena arise such self-focusing, self-modulation, mode-conversions, and waveguide envelop solitons can occur and affect the propagation of surface seismic waves.

3.7.3 Nonlinear surface waves of shear-horizontal polarization. Skimming waves

The Bleustein-Gulyaev waves are surface acoustic waves which, like (generalized) Rayleigh waves, propagate on a planar surface of a homogeneous piezoelectric medium and are therefore nondispersive. The way in which they are influenced by nonlinearity is, however, quite different from that of sagittally polarized surface waves. This is because of the fact that like Love waves (De 1970), the Bleustein-Gulyaev waves excite a second harmonic of sagittal polarization. Due to the change of polarization, this generation of the second harmonic is not resonant. However, there is a resonant interaction of the fundamental Bleustein-Gulyaev wave with its third harmonic which is again of shear-horizontal polarization and is accompanied by a nonzero electrostatic potential. This resonant interaction is either direct through third-order nonlinearity or indirect via second-order nonlinearity in a two-step process (Mayer 1991). Since rock is not piezoelectric we will not elaborate further this topic about Bleustein-Gulyaev surface waves. However, it is interesting to discuss nonlinear surface SH-waves in the case of nonpiezoelectric materials. We have seen surface SH-waves only in the case of Love waves which can be generated only in layered waveguide systems. Love wave are highly destructive seismic waves because of their SH-polarization, however they relatively rare in the seismologic practice because of the specific geological conditions that they need to be generate and propagate. However, if surface SH-waves could be generated on a regular half-space surface they would have a major impact during earthquakes.

The linearized equations of motion and boundary conditions for surface elastic waves do not admit solutions of surface acoustic waves of shear-horizontal polarization propagating on a planar homogeneous, nonpiezoelectric, isotropic medium. However, there is a bulk wave of shear-horizontal polarization propagating along the surface satisfying the traction-free boundary conditions at the surface called surface skimming bulk wave. The question is whether nonlinearity can lead to localization of this surface skimming bulk wave at the surface and makes it propagate as a surface elastic wave. This problem has first been investigated by Mozhaev (Mozhaev 1989) in an isotropic elastic medium. Assuming a displacement in the form:

$$u(\mathrm{x},t) = U_0(\mathrm{z})\left[qx - \omega t + \phi\right] \qquad (3.7.3.1)$$

Mozhaev derived $u_0(y)$ in the case a negative nonlinear coefficient α:

$$U(\mathrm{z}) = \mathrm{U}_0 \operatorname{sech}\left[q^2 U_0 \sqrt{-\frac{\alpha}{8}} z\right] \qquad (3.7.3.2)$$

and a dispersive relation:

$$\omega = \sqrt{\frac{\mu}{\rho}}q\left[1 + \frac{\alpha\left(qU_0\right)^2}{8}\right] \qquad (3.7.3.3)$$

Similar solutions of the equation of motion and boundary conditions for the displacement field have been obtained by other authors (Kosevich (1990) for shear-horizontal waves localized by nonlinearity at a planar interface between two homogeneous elastic media.

Skimming SH-waves located close to the surface due to nonlinearity and dispersion is a phenomenon that could have a great impact in seismology. To date only theoretical solutions have been found about the existence of skimming waves. Such waves have not been observed experimentally because of weak nonlinearity and dispersion in materials used in physical acoustics. However, strong nonlinearity and high dispersion in the rock media offer opportunity for experimental study of skimming waves and their impact as seismic SH-waves. Knowing how destructive seismic Love waves could be during earthquakes, skimming waves grab the attention because of much higher probability to be generated in normal geological conditions without the need of specific layers and waveguide structures as in the case of Love waves.

3.8 Nonlinear elastic waves in a dispersive medium

Indeed, as shown in Fig. 3.8.1 the crust's strength increases linearly in the elastic zone with predominantly brittle rocks and reaches a maximum at a certain depth where a brittle-ductile transition occurs followed by non-linear exponential decrease of the strength with depth.

Most of the elastic energy transport having an impact on the earthquake's effect happens in the brittle elastic zone down to the brittle-ductile transition zone. Farther down in the non-linear elastically ductile zone the attenuation of the bulk waves increases. This means that all surface acoustic waves that will emerge on the surface will carry energy density that depends essentially on the specific geological status of the brittle elastic zone.

We note that Fig. 3.8.1 gives only a general idea about the crust's strength variation with depth. In reality the profile of crust's strength can strongly vary from area to area. In average the maximum crust's strength is in depth zone between 13 km and 18 km but this certainly is not valid everywhere on the Earth. In many cases zones with linear elasticity are interrupted by mix zones of brittle-ductile rock modifying the crust at various depths and resulting in non-linear elasticity. The heterogeneous structure of the crust is a

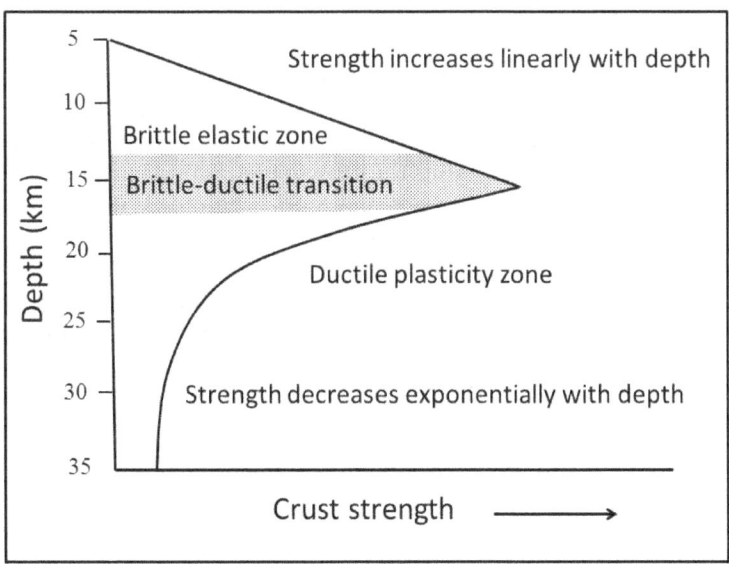

Fig. 3.8.1. The crust strength as a function of depth.

mix of domains with linear and non-linear elasticity. The energy transport is different in the linear and non-linear zones and it will vary depending on the specific geophysical profile of the crust. The lithosphere in general is highly dispersive continuum.

Nonlinear wave propagation in dispersive media differs from nonlinear propagation in nondispersive media. The growth of high harmonics cannot take place in dispersive media because they cannot couple resonantly between them and with the fundamental wave. In a medium with strong dispersion only the fundamental mode can propagate without any harmonic waves.

There is a significant difference between propagation of nonlinear elastic waves in nondispersive and dispersive media. In a nondispersive media high harmonics could couple to each other and to the fundamental mode in resonance conditions. In a dispersive media this coupling cannot happen. If the dispersion is strong the creation of high harmonics will be suppressed and it is possible to get sinusoidal solutions of the equation of motion if the nonlinearity is not very strong. These quasi-sinusoidal solutions are not necessarily stable. The theory of dispersive surface elastic waves is analogous to the theory of nonlinear electromagnetic waves propagating in optical waveguides. There are various mechanisms causing dispersion. The most relevant for seismic waves is the propagation of Rayleigh waves in nonlinear medium covered by a layer of a material with different elastic properties. The wavelength of the surface elastic wave $\lambda_{SAW} = 2\pi/k_{SAW}$ is assumed to be much larger than the thickness of the layer d. Maradudin and Mayer (Maradudin and Mayer 1990) have done the calculations assuming the layer is linear with a nonlinear dispersive Rayleigh wave with purely sagittal polarization as well as with Love wave with an SH-polarization. In the case of Rayleigh waves they have found that the dispersion inhibits the growth of a second harmonic. However, in the case of Love waves there is a second harmonic with polarization in the sagittal plane, perpendicular to the fundamental Love-mode which is SH-polarization and, therefore, there is no resonance between the second harmonic and the fundamental mode. The layer inhibits the growth of odd harmonics which would couple resonantly with the fundamental mode in the absence of the layer. The calculations demonstrate a direct contribution of the third-order nonlinearity and indirect contributions of the second-order nonlinearity via the second harmonic with strong compensations between elastic moduli of different order such as negative contributions of the second-order nonlinearity and the positive contribution of third-order nonlinearity. In the case of nonlinear Love waves a gradual variation of a beam profile along the propagation direction can occur when nonlinearity can counteract diffraction and give rise to the formation of stable channels—self-channeling or self-focusing. No experimental confirmation has been reported.

Laser techniques have been used for the generation of very high amplitude pulses with acoustic Mach numbers about 0.01 (Kolomenskii et al. 2003). In such cases it is useful to describe the SAW magnitude by a dimensionless acoustic Mach number $M = v/V_R$, where v is the amplitude of the surface velocity and V_R is the propagation velocity of the Rayleigh surface wave. Such waves drive the medium into the nonlinear elastic regime and shock fronts can be formed during their propagation. The time evolution of the waveform shape depends on the nonlinear acoustic parameters of the medium of propagation. Kolomenskii (Kolomenskii et al. 2003) have demonstrated that a compression of the high-amplitude SAW pulses takes place in stainless steel. A formation of a relatively stable portion of the waveform has been observed of soliton-like SAW pulses. The increase of the wave amplitude creates a possibility that the dispersion and nonlinearity balance each other in such a manner that the formation of a relatively stable portion of the wave can occur as a soliton (Lomonosov 2002).

Laser generation of nonlinear SAW pulses can result in strong nonlinear effects such as the formation of shock fronts and drastic changes of the pulse shape and duration (Lomonosov et al. 1999; Kolomenskii 1997). In Fig. 3.8.2 SAW pulses with different initial amplitudes are shown after propagating a distance of 40 mm. At higher amplitudes a relatively stable soliton-like pulse is formed in the tail of the waveform. The waveforms of the pulse at different distances are shown in Fig. 3.8.3. The sequence of the plots shows how the soliton develops.

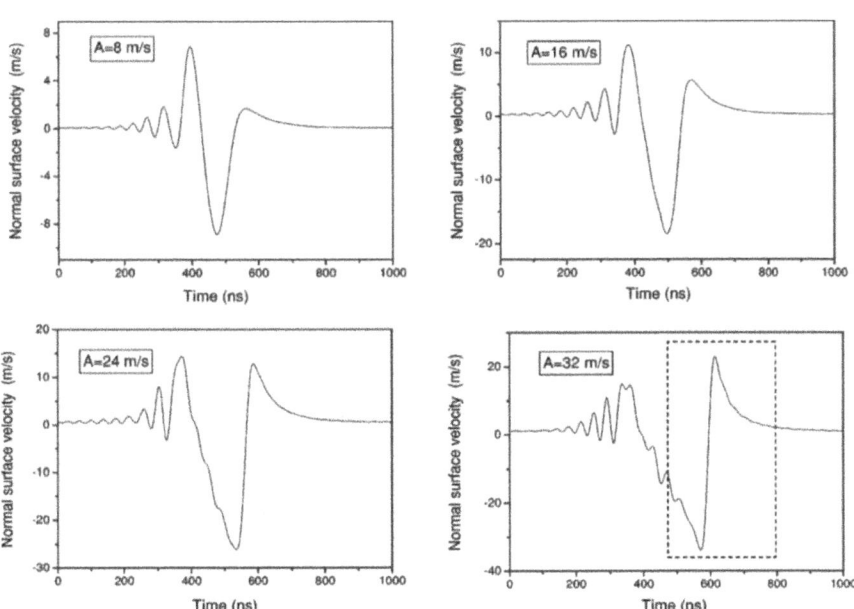

Fig. 3.8.2. A SAW pulse distorted by dispersion and nonlinearity. The formed soliton-like portion of the pulse is indicated by a dashed box (Kolomenskii 2003).

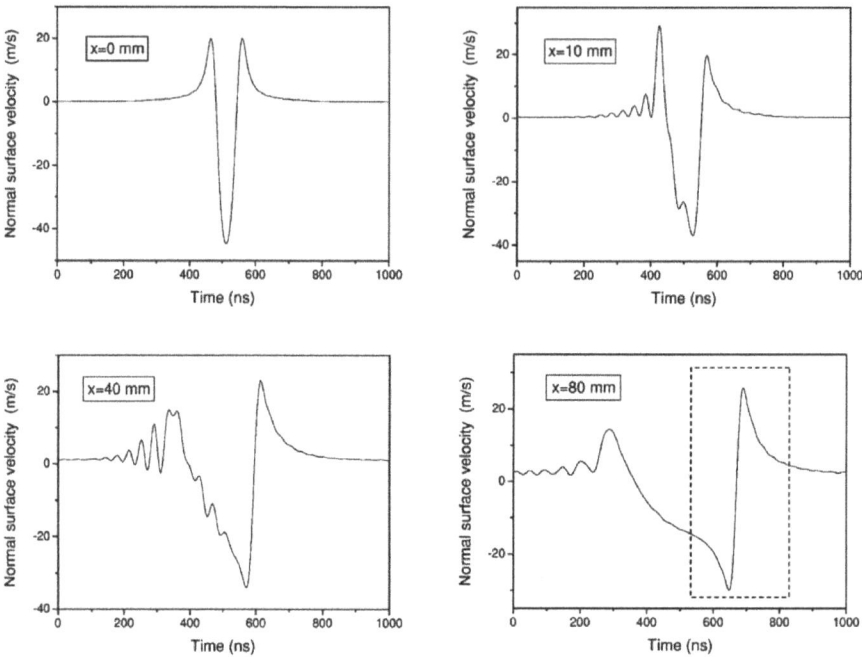

Fig. 3.8.3. A distorted SAW pulse at various distances. The formed soliton-like portion of the pulse is again indicated by a dashed box (Kolomenskii 2003).

3.8.1 Parametric amplification of Rayleigh waves

Weak elastic waves can be amplified by strong pump elastic waves of a higher frequency. The phenomenon is called parametric amplifications. It exists also in optics where a weak electromagnetic wave of frequency ω gets amplified through energy transfer from a strong pump wave of frequency 2ω. If there is no dispersion a great number of higher harmonics can grow. The energy of the strong pump wave will be distributed between them preventing in this way the amplification of the weak elastic wave.

In acoustoelectronics parametric amplification of a weak signal is a desirable phenomenon, so techniques have been proposed to prevent transfer of energy to higher harmonics by selectively damping them. It has been demonstrated that significant parametric amplification of body elastic waves can be achieved if the 3rd and 4th harmonics are attenuated (Andreev 1985).

It has been demonstrated that such high amplification can be achieved with Rayleigh surface waves (Maradudin 1990).

In contrast to acoustoelectronics, parametric amplification of seismic waves is a highly undesirable phenomenon. It is interesting to see how high the parametric amplification of bulk and Rayleigh waves can go in acoustoelectronics as a reference how scary this phenomenon is in the case of seismic waves.

Parametric amplification of Rayleigh waves has been further studied and ways for enhancing the parametric amplification have been proposed (Krilov 1993). Rayleigh propagating in a harmonic elastic material (only one nonzero Murnaghan nonlinear parameter is considered) has been considered using an infinite set of coupled nonlinear evolution equations (Parker 1988) modified by involving viscous energy dissipation $\sim \omega^2$ and dissipation of elastic energy from periodic surface gratings to calculate the maximum achievable values of parametric amplification in elastic media with and without dissipation.

Figure 3.8.1.1 shows the weak Raleigh wave amplitude (solid line) of frequency ω and the pump Rayleigh wave amplitude (dashed line) of frequency 2ω as functions of the distance X from the source for the two lower harmonics in presence of dissipation (denoted by index D) and in absence of dissipation (Krylov 1993). As a reference the distance of Rayleigh wave

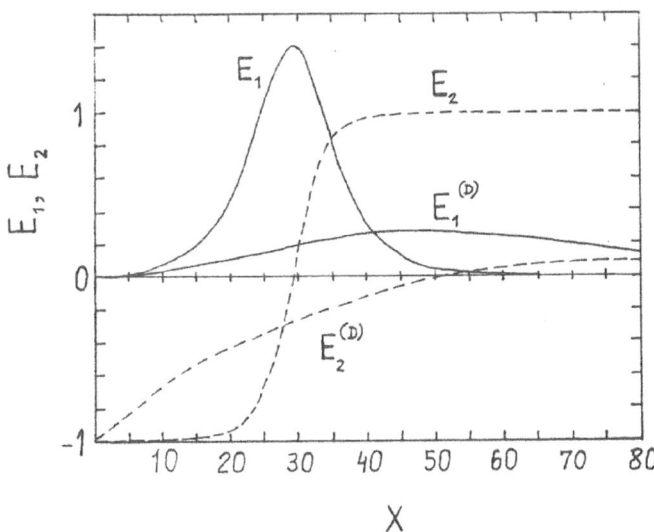

Fig. 3.8.1.1. The weak Raleigh wave amplitude (solid line) of frequency ω and the pump Rayleigh wave amplitude (dashed line) of frequency 2ω as functions of the distance X from the source for the two lower harmonics in presence of dissipation (denoted by index D) and in absence of dissipation (Krylov 1993).

shock formation is $X_{sh} = 3$. The amplitudes at $X = 0$ are $E_{10} = 0.01$ and $E_{20} = -1$, respectively. Without dissipation the amplitude E_1 grows until $X = 29.02$ where $E_{1,max} = 1.424226$. Then the second harmonic amplitude E_2 grows asymptotically to a value slightly above 1 while E_1 tens asymptotically to 0. In the presence of small dissipation ($P\omega^2 = 0.01$) the amplitude $E_1^{(D)}$ reaches maximum $E_1^{(D)} = 0.2997$ at a long distance of $X = 47.5$. The pump wave amplitude goes to zero short beyond $X = 47.5$. In the presence of dissipation the amplitude $E_2^{(D)}$ reached maximum at $X = 77.5$ and after that point tends to 0 together with $E_1^{(D)}$.

The above calculation involving two lowest harmonic have been extended into calculation of the maximum achievable coefficient parametric amplification K_{max} of Rayleigh waves. In the cases of absence of dissipation and in presence of dissipation K_{max} has been found to be equal to $K_{1max}/E_{10} = 141$ and $E_{1max}^{(D)}/E_{10}^{(D)} = 30$. In the usual conditions of acoustoelectronics experiments the propagation of nonlinear Rayleigh waves is accompanied of the growth of a great number of interacting harmonics, so K_{max} has not been achieved. It will be interesting to discuss the situation of nonlinear seismic Rayleigh waves propagating in highly dissipating, dispersive, and heterogeneous continuum of the lithosphere. Such a discussion will take place later in this Chapter.

Figure 3.8.1.2 shows $E_1^{(D)}(X)$ and $E_2^{(D)}(X)$ for twenty interacting harmonics. For $E_1^{(D)}(X)$ (solid curve A) the maximum value is only 0.03794 at $X = 4.5$ (close to the distance of Rayleigh wave shock formation). Beyond the point $X = 4.5$ the amplitude decreases because the energy of the pump wave is used for the generation of higher harmonics. In this case $K_{max} = 3.794$. This situation of small parametric amplification is similar to bulk elastic waves.

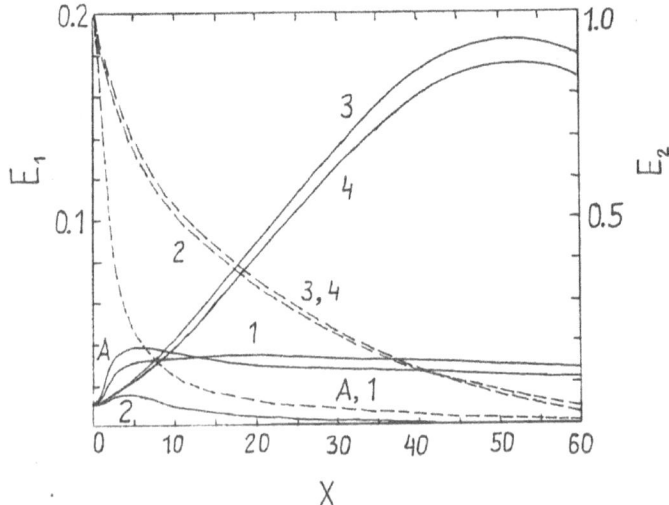

Fig. 3.8.1.2. The influence of the selective damping of the 3rd, 4th, and 5th harmonics (Krylov 1993).

It is interesting to follow Krylov's analysis on the influence of selective damping of the 3rd, 4th, and 5th harmonics shown in Fig. 3.8.1.2. The amplitude E_1 as a function of the distance X is shown in solid lines, while the amplitude E_2 is presented in dashed lines. Damping only the 3rd harmonic (curve 1) does not increase the parametric amplification much in comparison to absence of selective damping (curve A). However damping 3rd and 4th harmonics bring a maximum coefficient of parametric amplification K_{max} = 18.60 at X = 51.3. Damping the 5th harmonic together with 3rd and 4th does not increase any further K_{max}. Krylov propose a technique of damping 3rd and 4th harmonics to use Bragg diffraction grating on the surface of propagation that will provide selective reflection of these harmonics.

3.8.2 Solitary surface elastic waves

We have seen in Section 2.5.4.2 that in a nonlinear dispersive system, an initial disturbance can evolve into a bulk solitary elastic wave that retains its shape over a long distance. It has been found that when solitons collide with each other, they resume their initial wave forms and speeds. Bulk elastic soliton have been experimentally observed (Hao and Maris 2001). It has been stated that nonlinearity and dispersion are the main reasons bulk elastic solitons to exist. In this Section we will discuss solitary surface elastic waves, solitary elastic pulses, and surface elastic solitons as well as the conditions in which they are created. The difference between elastic solitons and solitary elastic pulses is that solitons survive elastic collisions with each other while solitary elastic pulses do not survive such collisions (Mayer 2008). Usually surface elastic waves are not dispersive waves except Love waves which propagate in a layer structures forming and elastic waveguides. Solitary acoustic pulses have been realized experimentally using pulsed laser excitation (Lomonosov and Hess 1999, Lomonosov et al. 2002). The dispersion has been generated artificially by covering the homogeneous elastic medium with a thin film made out of a suitable material to realize normal or anomalous dispersion. A comprehensive presentation of the theory of surface acoustic solitary waves has been given by Eckl et al. (Eckl et al. 1998a,b, 2004). The agreement between calculations based on nonlinear elasticity theory and experimental results on solitary pulse shapes has been demonstrated in these works. The measurements were performed by pulsed laser excitation of high-intensity acoustic pulses in layered systems. As a main result, the anisotropy of the substrate was found to have a strong influence on the pulse shapes. It has been shown in numerical simulations that these solitary pulses normally perform highly inelastic collisions between each other. Only for a Korteweg–de Vries (KdV) type linear dispersion law which is realized only in specific elastic mismatch between the substrate and a coating film solitary pulse collisions are nearly elastic. An important feature of surface acoustic solitary waves is their two-dimensional character. They have a nontrivial depth profile which may be constructed from their associated strain distribution at the surface. The latter can be determined from a one-dimensional evolution equation with strongly nonlocal second-order nonlinearity. The derivation of the evolution equation, as well as the reconstruction of the depth profile, was done with the help of asymptotic methods that have an approximate character and are valid for weak nonlinearity and weak dispersion. Eckl et al. (Eckl et al. 2004) have compared different variants of the theory provided in the literature were compared, and it has been shown explicitly that they lead to the same evolution equation. With these asymptotic methods stationary periodic solutions of the equations of nonlinear elasticity were constructed that are uniformly valid up to depths of the order of a typical wavelength divided by a typical strain. Due to the absence of material dispersion in the acoustics of solids, there are several physical systems where acoustic waves are nondispersive. By modification of the propagation geometry, dispersion can be introduced and tailored in a controlled way. One example is the system investigated by Eckl et al. of generalized Rayleigh waves propagating along the planar surface of a homogeneous elastic half-space. Normal and anomalous dispersion of the SAWs was realized by depositing a thin isotropic film onto the substrate. The nonlinearity in the corresponding evolution equations is partly of third order. An important difference is the scale invariance of the nonlinearity of homogeneous elastic media in the acoustics context, which poses a challenge for future investigations. Material dispersion in physical acoustic is weak or nonexistent for both bulk and surface elastic waves.

To study the process of generation and propagation of solitary elastic waves artificially controlled geometric dispersion systems have to be introduced in order to excite soliton waves. This is not the case of the heterogeneous body of the Earth where both material and geometric dispersion are normal phenomena. Rocks form a highly dispersive media which is very welcoming environment to body and surface elastic solitons. If Rayleigh solitons can be excited on the surfaces of materials used in physical acoustics only by introducing layers with matching elastic parameters, the highly dispersive Earth's surface offers plenty of possibilities for the generation and propagation of generalized Rayleigh solitary waves. Geometrical dispersion allows Love solitons to propagate as well, however, as we have seen these waves require a layered elastic waveguide system to propagate which is relatively rare in the usual geological structures of the Earth. No such a restriction exists for Rayleigh waves making surface seismic solitons the most dangerous waves to man-made constructions. We have discussed the process of generation of Rayleigh waves from upcoming to the free-surface body waves and vice versa—conversion of Rayleigh waves into body waves in specific geometric characteristics of the surface. This means that solitary body waves and solitary Rayleigh waves can freely interact with each other in the dispersive rock. This creates a unique environment to study elastic solitons which does not exist in physical acoustics. No such research has been done yet on seismic waves. A comprehensive study of the processes of generation and propagation of seismic solitary waves certainly will help to understand better earthquakes and will allow more efficient defense against their destructive power to be developed.

Surface waves propagating on shallow water constituted the first system in which solitary waves have been observed and studied, and they continue to play an important role in the field of nonlinear waves. As a result of their long wave lengths, tsunamis behave as shallow-water waves. A wave becomes a shallow-water wave when the ratio between the water depth and its wave length gets very small.

Korteweg–de Vries equation (KdV equation) was derived as a nonlinear evolution equation for surface shallow-water waves. The question was whether solitons may also propagate along the surface of solids. Several experiments with high-intensity Rayleigh waves have revealed features of the wave form evolution that are very reminiscent of the soliton dynamics of the KdV equation. For a linear Rayleigh waves the evolution equation (Eq. 3.7.1.13) in the absence of dispersion, integrating by parts and using the boundary conditions at $z \to 0$ and $z \to -\infty$ leads at second order of ε to the evolution equation for the strain amplitudes:

$$i \frac{\partial B_k}{\partial \tau} = k \left[\sum_{0<k'<k} F(k'/k) B_{k'} B_{k-k'} + \right.$$

$$\left. \sum_{k>k'} F(k'/k) F^*(k'/k) B_{k'} B^*_{k'-k} \right] + \alpha_R k^2 B_k \tag{3.8.2.1}$$

After Fourier transform with respect to the spatial coordinate for $k > 0$ Eq. 3.8.2.1 leads to the Korteweg–de Vries equation for $m = 3$:

$$i \frac{\partial B_k}{\partial \tau} = k\Phi \left[\sum_{0<k'<k} B_{k'} B_{k-k'} + 2 \sum_{k'>k} B_{k'} B^*_{k'-k} \right] + \alpha k^m B_k \tag{3.8.2.2}$$

For $m = 2$ Eq. 3.8.2.2 represents the equation of Benjamin-Ono. Both equations can represent evolution equations of a nonlinear surface elastic wave.

In Fig. 3.8.2.1 two examples compared with corresponding localized solutions of the KdV and the BO equations. In the limit $\zeta \to 0$ they represent a one-parameter family of solitary wave solutions having the same relations between velocity, width, and peak height as the solitons of the BO equation. A characteristic feature is the "Mexican hat" shape with two local minima, which distinguishes them from the KdV and BO solitons.

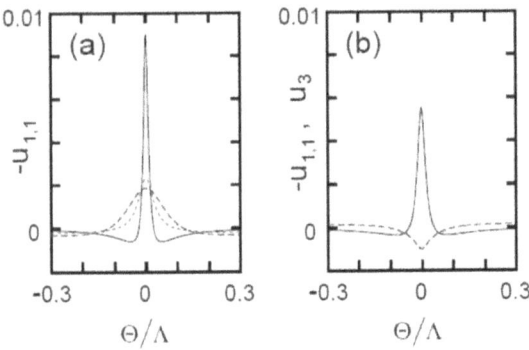

Fig. 3.8.2.1. (a) The strain component $-u_{1,1}$ at the surface associated with a solitary pulse train solution of Eq. 3.8.2.1 for Si(001) (100), $\zeta = 0.3$ (solid). Solitons of BO equation (dotted) and KdV equation (dashed) are presented. (b) Strain component $-u_{1,1}$ (solid) and surface elevation profile (u_3 at $z = 0$, dashed) corresponding to a Rayleigh solitary wave with z - 0.5. (Eckl and Mayer 1998).

Eckl and Mayer found an unusual consequence of the specific nonlocality of the nonlinearity. The fact that the asymptotic behavior of the solitary pulses does not need to be governed by the linearized version of the evolution equation means that it may also be influenced by the nonlinear terms. This may be demonstrated by choosing the function F in the evolution equation (Eq. 3.8.2.1) to be a real constant and the last term on the right-hand side to be replaced by $-ak^3 B_k$. This dispersion term appers when the acoustic mismatch between film and substrate is such that the coefficient α_R vanishes. At the limit $\Lambda \to \infty$ (large wavelength) an analytic solitary wave solution has been found to have a Mexican hat strain profile at the surface solution has been found to have a Mexican hat strain profile at the surface solution has been found to have a Mexican hat strain profile at the surface $z = 0$:

$$u_{1,1} \propto \frac{2\beta^2}{\left(\beta^2 + X^2\right)^2} - \frac{1}{\beta^2 + X^2} \qquad (3.8.2.3)$$

where $\beta^2 = 3\alpha_k/k$ and $X = (1 + k)x - V_R t$. k is a free parameter. It has been demonstrated that the solitary wave solutions of Eq. 3.8.2.1 for nonlinear Rayleigh waves are stable.

It has been mentioned that a necessary condition for a solitary wave to be called soliton is the property of its collision with other solitary waves. If it emerges out of a collision as a stable pulse with no additional radiation (elastic collision) this solitary wave is a soliton. Without dispersion the solitary surface waves cannot survive collisions with each other. Dispersion is a mandatory condition for solitary surface waves to exist. In the physical acoustics experiments such dispersion is introduced artificially by coating the substrate with a thin film or corrugating its surface. Since seismic surface waves are nonlinear and propagate on heterogeneous and dispersive surfaces all these results obtained by studying solitary acoustic waves suggest that stable seismic solitons could exist and, therefore, research in this field is a must.

4

Modelling and Earthquake-Resistant Design

4.1 Statement of the problem

From the analyses of linear and nonlinear propagation of bulk and surface elastic waves in isotropic, anisotropic, nondispersive, dispersive, and heterogeneous media we can draw some conclusions about haw theoretical and experimental results of physical acoustics can be applied for modelling seismic-wave deflecting, mode-converting, absorbing, or scattering shielding systems made of methamaterials.

i) Body and surface seismic waves are nonlinear finite-amplitude elastic waves that propagate in the nonlinear, dispersive, and heterogeneous continuum of the Earth. The destructiveness of an earthquake depends on the type of the seismic waves reaching man-made constructions and their power, but mostly on the rock nonlinearity and dispersion. Body P- and S-waves coming from the bulk of the Earth to the its surface can get converted into surface waves and vice versa. Raleigh, skimming, and Love surface seismic waves have the strongest impact on man-made constructions. Direct body seismic waves can also cause damages, however, the probability for them to reach directly a city is much lower than the probability to get converted into surface waves, so their impact should be considered mainly as a power source of surface seismic waves.

ii) Most of the seismic activities that have an impact to man-made structures happen in the lithosphere. The upper part of the lithosphere—the crust—has mostly a brittle structure all the way down to a brittle-ductile transition point at about 15 km depth followed by a ductile structure reaching the upper mantle where the hard crustal olivine rock brings back the brittle structure which continues down to another brittle-ductile transition point about 40 km deep. The elastic properties of brittle zones are mostly linear, so nonlinearity is due predominantly to nonlinear finite-amplitude seismic wave propagation. Dispersion of the heterogeneous rock mass together with the nonlinear propagation of the seismic waves result in growth of higher harmonics and waves with combination frequencies, phenomena such as parametric amplification, self-modulation, self-focusing, solitary wave propagation, solitons, wave-wave interactions, and skimming waves generation.

iii) Linear elasticity theory cannot be used to describe the propagation of seismic waves because in most cases terms of second and third order in the Taylor's series development of the seismic wave strain cannot be neglected. The theory of material linear elasticity and linear elastic wave propagation is well developed and well understood. This theory can be used in cases of weak nonlinearity and weak dispersion do describe certain phenomena, however, seismic waves are always nonlinear waves that obey nonlinear state equations. In a linear system the waveform does not change, it remains invariant during the propagation of the wave. In a nonlinear wave propagation the waveform evolves during the propagation because different parts of it move at different velocities. Linearizing these equations leads to relatively simple exact solutions, but there is a risk to omit important phenomena that in the case of seismic waves are crucial to their generation and propagation. Earthquakes happen only in a nonlinear world and they wouldn't exist would the world were linear. Therefore, as with all approximations it is necessary to evaluate its validity and range of errors.

iv) The theory of nonlinear elasticity and nonlinear wave propagation is much closer to reality that the linear elastic theory and linear elastic wave propagation. Nonlinear elasticity has been researched extensively and many cases of quadratic nonlinearity have been well developed especially with the help of computer simulation and numerical solving governing equations. However, nonlinear elastic wave propagation is still in process of development and many phenomena have not been fully understood. Introducing cubic or higher order of nonlinearity increases exponentially the complexity of the problem. In many cases third-order stiffness constants are much smaller than second-order constants which simplifies the problem. However, this is not the case of many rocks where cubic. Nonlinearity is stronger than the quadratic nonlinearity. Stress-induced anisotropy of rocks can be caused by pockets of higher internal stress that can be releasing when a seismic wave passes near by, but also a high-amplitude wave. Can cause stress-induced anisotropy. In summary nonlinearity is due to two different causes: 1) nonlinear strain-stress relationship of the media of propagation, and 2) finite-amplitude elastic wave that propagates in a nonlinear way even in linear media, i.e., the waveform evolves in time. Close to the hypocenter of the earthquake a finite-amplitude wave may have sinusoidal waveform, but because of the different values of the velocity of propagation at different portions of the wave, with increasing the distance traveled by the wave its waveform gets distorted. In low dissipation media during the evolution of the waveform the wave's crests overtake the wave's troughs leading to a shock wave.

v) Dispersion is another important factor. An elastic wave that propagates in a medium with low dispersion grows high order of harmonics that can interact with each other as well as with the fundamental wave. The fundamental mode transfers its energy to higher harmonics and eventually decays. However, if the third and fourth harmonics are dissipated phenomena of parametric amplification can occur and a strong second harmonic Raleigh wave can 'pump' energy into the weak fundamental mode amplifying it in some cases to more than two orders of magnitudes. If the medium of propagation is highly dispersive the situation is completely different. The higher harmonics cannot interact with each other and the fundamental mode can keep its energy and travel long distances in low attenuation media. This mode is not necessarily stable, especially if the elastic wave is weak. However, a high-amplitude elastic wave propagating in materials with specific elastic properties such as rocks can be very stable and can travel long distances. There are two types of dispersion: 1) material dispersion caused by scattering and energy dissipation, and 2) geometrical dispersion caused by the existence of specific boundary conditions in the medium of propagation forming waveguides as is the case of Love waves. In the presence of strong dispersion 'anomalous' nonlinear phenomena such as self-focusing, self-modulation, and elastic solitons can occur. Often on seismograms P-, S-, LR-, or LQ-waves are identified from the time of their arrival at seismic stations knowing that longitudinal waves are faster than shear waves which are faster than surface waves. In many seismograms this is what is observed, but in many others the velocity rule does not work. This happens because some seismic waves reaching the seismometer do not come directly from the focus of the earthquake; there are generated through mode conversion somewhere between the focus and the point of recording. This is especially valid in the case of surface seismic waves which are never direct waves; they are always generated from body waves. Seismograms registering vertical, transverse, or radial components of the seismic displacement provide more information about the type of the detected waves that those registering just the vertical amplitude of the strain. Even more informative are seismograms recording velocity and acceleration of strain in various directions. Carrier waves which are part of the diffuse seismic field (coda) get modulated or overmodulated by lower frequency body or surface waves that often are solitary waves or solitary pulses with typical shape of 'Mexican hat' of velocity of displacement. The new wave results from a multiplication of carrier and modulating waves and often has greater amplitude than its components. This new wave gets registered by the seismometer. The amplitude modulation can occur only because of nonlinearity of the propagation. Growth of waves of combination frequencies results in 'seismic beat'.

vi) Heterogeneity of the medium of propagation is the most important factor that shapes the propagation of elastic waves in rocks. It causes nonlinear effects, anisotropy, and strong dispersion. Because of dispersion there are no interactions between higher harmonics and, therefore, they vanish pretty fast. Weak elastic waves vanish fast as well because of intense scattering, but they have low impact during earthquakes anyway. However, strong high-amplitude waves can become very stable and can travel long distances keeping their high energy density. Their impact during earthquakes can be significant because they form the set of carrier and modulating waves. This is especially relevant in the case of Raleigh, skimming, and Love waves which have the highest impact of all types of seismic waves. Rock slabs in the Earth's crust form layered waveguide structures that can enhance strongly the dispersion and set up situations suitable for the formation of Love waves and surface solitons. However, such geological formation are not likely to be met often. In contrary rock nonlinearity and dispersion can promote often growth of stable skimming waves with Love-wave-like polarization and same level of destructiveness. In 'weak' alluvial soil composed of silt and clay scattering, refraction, and attenuation phenomena strongly increase dispersion resulting in amplification, self-focusing, trapping, and generating Raleigh-like surface waves at the basin edges that propagate at 100–200 m/sec. In the seismic wave normal frequency band these values correspond to shorter wavelengths that those of body P- and S-waves or Raleigh waves on rocks. Wavelengths of 50–100 m are in the range of normal city building dimensions which makes them suitable of causing standing waves and resonance effects.

vii) From the properties of nonlinear propagation of seismic waves it is possible to make a conclusion that designing and building seismic shield or cloaking systems around buildings or cities able to deflect or attenuate upcoming seismic waves during an earthquake is an attractive option for developing earthquake-resistant design and technology. Taking into account the complexity of the nonlinear seismic wave propagation it is certain that designing and implementing such deflecting systems is not easy, but it is possible, especially if methamaterials are used. Recent development of metamaterials science provides new direction for controlling seismic waves. Designs of cloaking systems for controlling bending waves propagating in isotropic heterogeneous thin plates have been proposed (Farhat et al. 2009). Seismic waveguide of metamaterials has been proposed for earthquake-resistant design to support conventional a seismic systems using metamaterials (Kim and Das 2012). The device is an attenuator of seismic waves based on a cylindrical shell-type waveguide composed of a great number of Helmholz resonators that create a stop-band in the seismic frequency range converting the seismic energy into sound and heat.

viii) Most of the difficulties come from nonlinear governing equations of state that do not have exact solutions and only numerical solutions can be achieved with appropriate approximations. At its present stage the theory of nonlinear elasticity and nonlinear elastic wave propagation cannot be of great practical help to civil engineers looking for real solutions to protect cities from seismic devastation. However, as many scientists have done throughout the history of the development of the nonlinear wave propagation theory, when facing unsurmountable mathematical problems using experimental methods to measure important parameters that cannot be calculated proves to be very helpful. Recently carried out mixed experimental-simulation experiments seem to be a prominent technique. Inserting realistic input values is a mandatory request for carrying out reliable simulations. Since soils possess various uncertainties providing realistic input values for the simulations requires *in situ* preliminary tests to adjust the soil parameters, such as shear modulus and quality factor. A test zone with thick homogeneousilty clay soil was selected near the alpine city of Grenoble (France) to carry out the experiment. The experimental setup and results are is shown in Fig. 4.1.1 (Brûlé et al. 2014). The depth of the basin with similar deposits is up to 200 m. The Rayleigh wave velocity was first estimated to be 78 m/s by the preliminary seismic

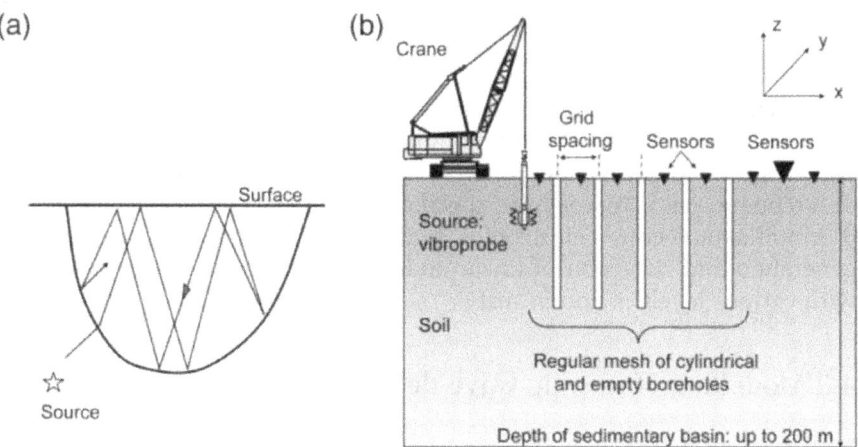

Fig. 4.1.1. (a) Seismic waves in an alluvium basin; (b) Seismic testing device (Brûlé et al. 2014).

test by pointing the wave time arrivalat various offsets from the source. Numerical simulations with finite elements were performed to predict the stop-band frequency for elastic surface waves. Since the seismic metamaterial test is challenging to model in full the three-dimensional Navier equations in unbounded heterogeneous media an asymptotic model which captures the wave physics at the air-soil interface only was used following approximate Mindlin plate model for time-harmonic surface flexural waves (Mindlin 1954).

In order to implement the above listed conclusions in (i)–(viii) we will discuss now practical solutions of physical acoustics experimental modeling that can be used in the design of earthquake-resistant systems for protecting man-made structures. Carrying out large-scale experiments in natural conditions is certainly closer to reality, but they are much more expensive and offer fewer options than laboratory experiments using small samples to improvise earthquake conditions. The focus will be on seismic-wave deflecting and attenuating systems as well as systems that prevent resonance phenomena. Taking into account the high level of seismic power density registered during many powerful earthquakes such as the 1985 Mexico City earthquake it seems that redirecting (or converting and redirection) the seismic waves away from cities or installing metamaterials preventing the occurrence of resonance phenomena offers better chances than absorbing or attenuating the seismic waves unless efficient techniques and metamaterials are used to convert the seismic energy into some other type of energy. Since surface seismic waves have the highest impact on man-made structures during earthquakes the focus will be put on their experimental generation, detection, refraction, scattering, mode-conversion, and attenuation. Body waves will be considered as well mostly as a power source of generating surface waves and also as an efficient way to drain out the seismic energy of surface waves by converting them back into body waves directed down to the Earth's bulk.

The experimental simulation samples can be isotropic, anisotropic, homogeneous, heterogeneous, nondispersive, and dispersive. Theoretical simulation can be carried out in parallel, however, the main attention here will be given to experimental modeling of systems able to deflect, scatter, or attenuate elastic waves from specific areas of the sample. Acoustooptics, thermoplastic, acoustoelectronics, and laser techniques will be used for generation and detection of linear and nonlinear elastic waves in samples with various levels of nonlinearity.

4.2 Modelling of seismic wave deflecting systems

Usually earthquakes last only a couple of seconds. There could be, of course, aftershocks, but the main shock is short as shown in Figs. 4.2.1 and 4.2.2 in

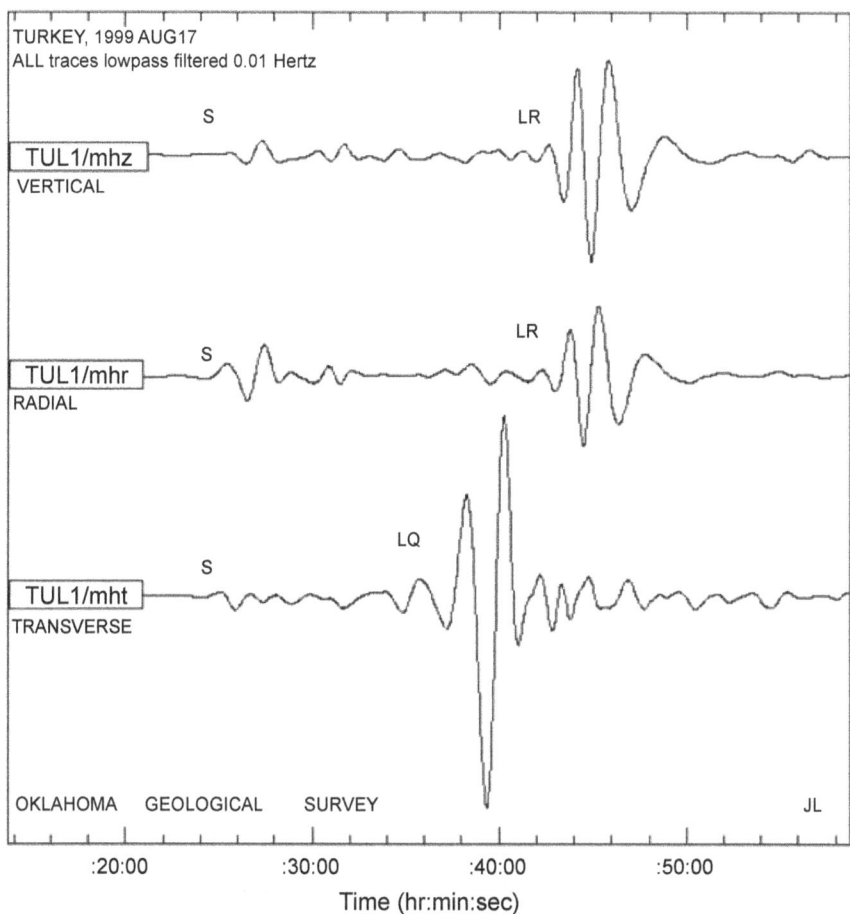

Fig. 4.2.1. The 1999 Turkey earthquake ground velocity recorded by OGS in vertical, radial (toward the epicenter), and transverse (at right angles to the direction of the epicenter) directions..

the case of a 1999 earthquake in Turkey and an explosion generated shock pulse in Minnesota. LR and LQ denote Rayleigh and Love waves, respectively.

We have shown that seismic waves are elastic waves that propagate in the heterogeneous lithosphere; however, it would be more accurate to speak about elastic pulses. Physically there is no difference in the propagation of an elastic wave and an elastic pulse. While the term 'elastic wave' suggests continuous or long lasting propagation phenomenon, the term 'pulse' suggests just a spurt of wave, or a short section of it. The correct term is probably 'pulse wave'. One can say that this is a matter of semantics, but actually is more than that, especially in the case of nonlinear propagation. The propagation velocity of body elastic waves or Rayleigh-type surface waves

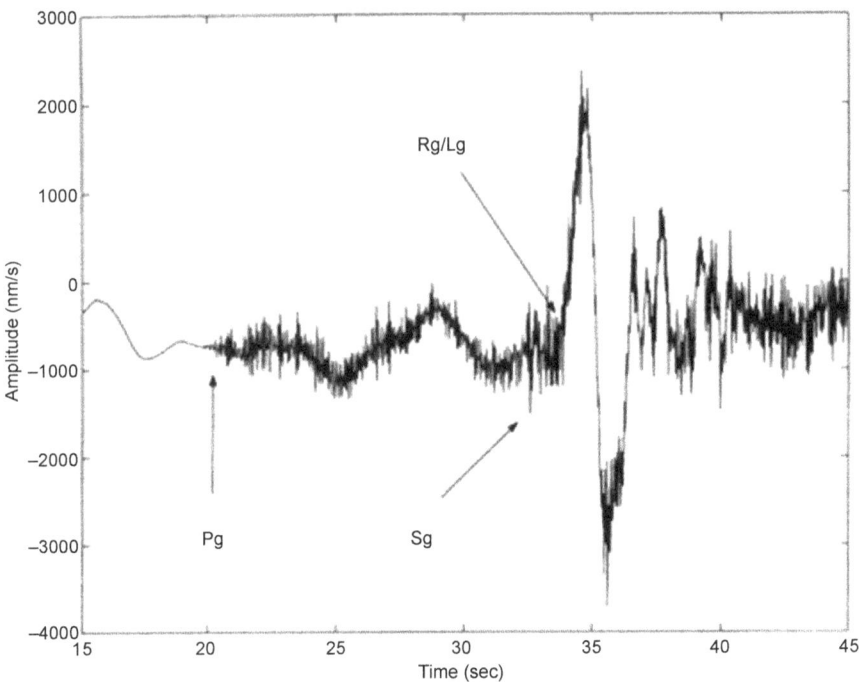

Fig. 4.2.2. Fired Explosions at the Minntac Iron Mine, Minnesota.

with small amplitudes propagating in homogeneous solid does not depend on frequency or the amplitude, i.e., there is no dispersion or nonlinearity. If also the attenuation in the medium of propagation is low a plane-front waves will retain their shape and amplitude during its travel. For elastic waves with large amplitudes, however, the velocity of propagation depends on the velocity of particle displacement of the medium and different parts of the wave profile move with different velocities causing changes in the pulse waveform. As the pulse progresses in time it waveform will change due to the wave nonlinear propagation. The pulse waveform evolution in time depends on nonlinearity strength. In the case of weak nonlinearity the waveform evolution is differs from the waveform evolution if the nonlinearity is strong. To understand the difference let us consider laser-generated surface elastic waves propagating on the half space of a homogeneous isotropic solid.

The elastic wave amplitude depends on the power of the pump laser, the thermo-elastic properties of the substrate, and the efficiency of the light-to-heat conversion. In the next sections special attention will be given to experimental modeling and simulation of earthquakes in small laboratory samples using powerful laser beams for generation of nonlinear bulk and surface elastic wave pulses (Kolomenskii et al. 1997; Lomonosov et al. 1999a, 2002b; Kumon et al. 1998). Figure 4.2.3 shows the low-passed filtered data from the August 17, 1999, Ismit, Turkey earthquake (Fig. 4.2.1) (right) and the vertical and transverse velocities measured at 5 mm from the source of a solitary surface elastic pulse generated by a 50 mJ Nd:YAG laser on a Si crystal (Kumon et al. 1998) as a comparison to the dashed-line marked area of the seismic pulse vertical velocity. The two curves are very alike at that specific time moment of waveform evolution exhibiting strong nonlinearity. However, when comparing acoustic and seismic waves it is important to take into account the level of nonlinearity because the pulse waveform evolution if very different at weak and strong nonlinearity. At higher nonlinearity the phase shifts caused by the interaction between different spectral components are comparable to the influence of dispersion, and under certain conditions these two processes may essentially compensate each other, as found for the pulse shown in Fig. 4.2.4 (Lomonosov and Hess 1999). The gradual frequency change in the oscillatory pulse essentially disappears and the pulse shape resembles two serial 'Mexican hats' (Eckl et al. 1998).

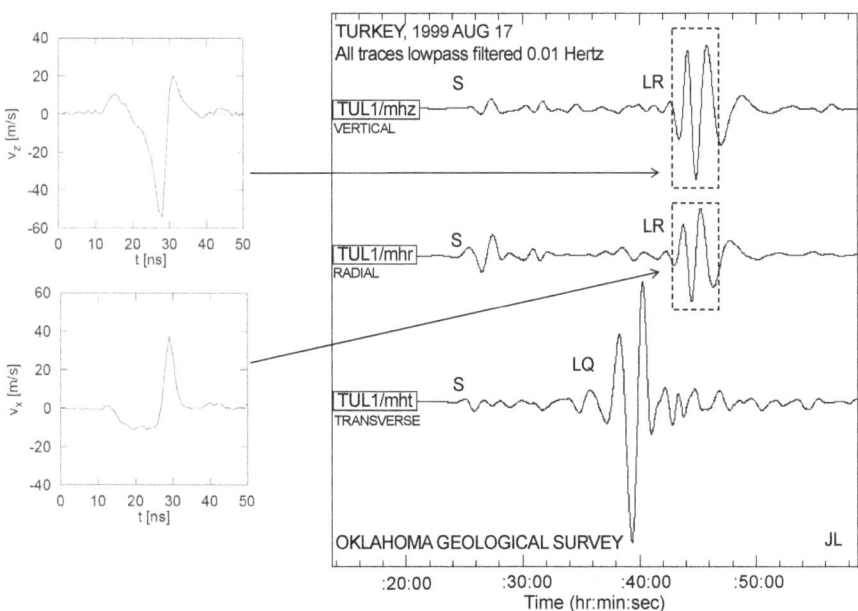

Fig. 4.2.3. Left—Vertical velocity (above) and radial velocity (below) measured at 5 mm from the source of a solitary surface elastic pulse generated by a 50 mJ Nd: YAG laser on a Si crystal (Kumon et al. 1998); Right—The 1999 Turkey earthquake ground velocity recorded by OGS in vertical, radial, and transverse directions (Fig. 4.2.1).

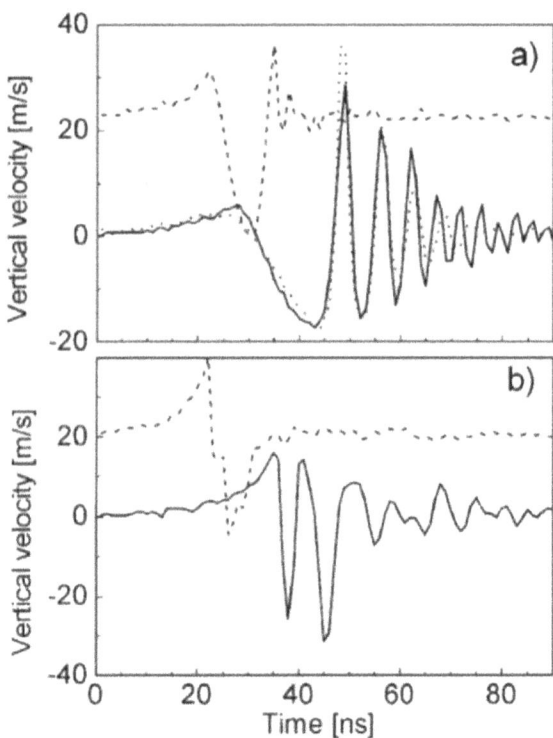

Fig. 4.2.4. (a) Waveforms of the vertical velocity measures in <112> direction of oxide coated Si (111) plane at 4 mm (dashed line) and 19 mm (solid line) and simulation at 19 mm (dotted line); (b) Waveforms measured for stronger laser excitation (Lomonosov and Hess 1999).

The above example shows that if we want the elastic wave pulses excited in a laboratory sample by a laser beam to propagate in a similar way as seismic wave pulses propagate we have to make sure that the laser power is high enough. We have considered so far nonlinear propagation in isotropic, homogenous, nondispersive, and low attenuation medium of propagation. Real seismic pulses propagate in high-attenuation, dispersive, and heterogeneous medium. In these conditions a weak pulse usually loses it power and gets dissipated quickly, but in some cases it can evolve into solitary wave pulse that retains its waveform over a long distance. This is due to the dispersion and cannot occur for nondispersive waves. High-amplitude pulses propagating in a heterogeneous and dispersive continuum add their nonlinearity to the nonlinearity of the medium enhancing various phenomena resulting from nonlinearity. Heterogeneity, elastic power absorption, and strong scattering make soil a strongly dispersive medium of propagation. The nonlinear dispersive propagation of seismic waves rases the question about the relevance of identifying P-, S-, LR, and LQ-waves on the seismograms as shown in Figs. 4.2.1 and 4.2.2 on the basis of their time of arrival at the seismic station. Self-modulation can occur anywhere between the epicenter and the seismic station and an S-wave can arrive faster than a P-wave. Comparing Figs. 4.2.3 and 4.2.4 where the vertical velocities have been registered a similarity with a 'Mexican hat' can be made which is typical for solitary pulse propagation or even a soliton wave.

Raleigh-like waves propagating on the half space of a homogeneous isotropic medium are nondispersive. However, if the half-space is covered with layers they become dispersive. Since the upper layer of the crust has mostly multilayer structure all Raleigh-like surface wave pulses will be dispersive. Love waves are always dispersive by default. Love waves require very specific boundary conditions with a thin rock layer where the velocity of propagation of the horizontally (parallel to the boundary) polarization SH-wave is slower than the velocity of propagation of the SH-waves in the substrate. However, when such boundary conditions are present, Love seismic waves can be very destructive to cities because they propagate in the rock layer as in an elastic waveguide where high level of seismic energy is confined. Therefore it is important to perform thorough geological studies around cities that are located in regions with a history of Love type earthquake to find out if rock layers suitable to accommodate Love waves exist. To eliminate any possible danger of appearance of Love waves during eventual earthquake it would be probably sufficient just to damage the waveguide structure using ground explosions. However, in nonlinear and dispersive rocks which is typical for the lithosphere skimming waves can be generated at a much higher rate than Love waves. Skimming waves do not need layered waveguide structures to propagate. Having Love-wave type of polarization and similar power of destructions skimming waves are much more difficult to detect by studying the geological structure of some region. In this situation it is much more efficient solution to develop seismic shields that can deflect the waves away from a city or mode-convertors built of metamaterials that can convert surface waves into bulk waves and send back down into the Earth's body.

Stonley waves are very similar to Love waves. They can be created only in certain boundary conditions between two rock layers with specific elastic properties allowing the propagation of two Rayleigh waves on both sides of the boundary. By eliminating such boundary conditions by destroying the interface the danger of having Stonley waves created during an earthquake could be minimized or fully eliminated.

The most often threat to cities during an earthquake represents the Rayleigh waves they could be formed on the Earth's surface easily from the diffuse seismic field in the crust. We have seen that the Rayleigh surface elastic waves have one quasi-longitudinal qP-component and one vertically polarized quasi-shear qVS-wave. The two waves are propagating together with a phase shift which in the case of isotropic medium is $\pi/2$. The qVS-wave makes the particles to move on quasi-elliptical patterns perpendicularly to the ground. Because Rayleigh wave can be formed on any free boundary it is difficult to predict their eventual epicenter which makes the task to eliminate them as a danger to cities more complicated. However, we have seen that Rayleigh waves can be reflected pretty efficiently by edges or be mode-converted into bulk elastic waves that will go back deep into the ground.

To protect a city from the devastating power of an earthquake it is necessary to 1) do geological study around the city and eliminate boundary interfaces in the rock mass that could facilitate the creation of Love and Stonley waves, and 2) design and build constructions for reflecting, attenuating, or mode-converting Rayleigh waves. The first point is straightforward; the second one is more complicated. The difficulties come from the fact that the Rayleigh seismic waves are generated and propagate in heterogeneous media. The theoretical study of such elastic waves is very complicated so is the design of systems that are able to deflect the seismic power away of the city is difficult. The old physical sciences technique when it is hard to calculate some parameter using mathematical methods then it is better to try to measure it probably could be applied to this challenging problem.

Building laboratory samples with heterogeneous structures for imitating artificial earthquakes is an attractive option. The only difference with a real scenario would be the frequency (or the wavelength) because the velocities of propagation could be adjusted easily to be close to the velocities of propagation of elastic waves in the ground. A diffuse elastic field can be generated in the sample using piezoelectric transducers as shown in previous section. Using wedged transducers as well as transducers made of specific cut of quartz or lithium niobate can be used for generating both P- and S-modes. To create Rayleigh surface waves on the surface of the sample interdigital grating electrodes can be used. Basically all techniques used in physical acoustic to generate and detect elastic waves with various polarizations can be used in this case to generate and detect qP- and qS-waves forming a diffuse 'seismic' field in the sample. Also some of the techniques used in physical acoustics to deflect, reflect, diffract, or attenuate bulk and surface acoustic waves can be used but more as a guidance than to design real systems for preventing seismic energy from reaching a city because of the more complicated propagation in heterogeneous media. In heterogeneous medium the design of such preventive systems can be done more efficiently by modelling the systems in our laboratory heterogeneous sample. Once the Rayleigh waves are created in the sample it would be necessary to install mock reflectors or attenuators in the way of the Rayleigh waves and measure experimentally their effect by measuring the output Rayleigh waves after they have passed through the mock constructions. Useful techniques for doing that can be borrowed from the non-destructive testing of materials—a section of physical acoustics that study mechanical, electric, magnetic, optical and so on properties of materials.

One of the most used techniques to study the generation and propagation of elastic waves in solids is light diffraction of elastic waves. When a monochromatic collimated beam passes through a grating it gets diffracted. This is called laser probe in physical acoustics. Such effect can be observed when a laser beam goes through a small aperture. The result can be observed on a screen as a multitude of bright spots representing diffraction patterns. Similar effect occurs when a laser beam crosses and elastic wave that propagates in an optically transparent medium. The elastic wave creates of grating of refraction indexes by modulating the density of the medium. The density (mass per unit of volume) depends on elastic strain and as we know the elastic wave cause a periodic strain in the medium of propagation.

The heterogeneous ground medium can be considered as composed by a great number of blocs each with its own elastic properties and mass densities. The size of the blocs are of the order of the elastic wavelength or larger. There is no need to take into account blocs that are smaller than the elastic wavelength because the elastic wave cannot "see" such blocs. During the earthquake elastic waves created as a result of multiple reflections and refractions from blocs' boundaries propagate in all directions. Thus a seismic diffuse field is created. Such diffuse elastic fields are created inside individual blocs as a result of internal reflections. In general such diffuse elastic field is week. It could shake city's building without causing construction damages. Only powerful elastic waves generated as a result of resonance effects in the diffuse seismic field that are directed towards the city as a bulk wave or as a surface waves present significant danger to the buildings and city's infrastructure. If appropriate reflectors and absorbers are constructed under and around cities located in earthquake areas they will deflect the bulk and surface seismic waves away from the constructions and diffuse their power. It is impossible to predict the type of an earthquake, its power, and the wavelength of the elastic waves that will attack the city. To be efficient in protecting the city from upcoming seismic waves the reflecting and absorbing constructions need to be able to operate in broad spectral ranges. Shapes, critical angles, and depth in the ground are critical parameters that have to be determined for each specific case. In the following chapters of this book we will focus on the design of such reflectors and absorbers.

4.3 Generation and detection of elastic waves

In the process of search for more efficient methods to prevent damages caused by earthquakes scientist have developed techniques to create artificial mock earthquakes using underground explosions or mechanical systems generating powerful vibrations in the ground under or around experimental buildings. A large number of sensors are used to detect various parameters of the artificially induced elastic waves and their impact. The research is costly; it requires massive installations, heavy equipment, and a significant deal of human resources.

In this chapter we will demonstrate that by using techniques developed in physical acoustics and acoustoelectronics micro-earthquake seismic waves can be generated and detected in laboratory samples smaller than $10^{-3}\,\mathrm{m}^3$ imitating mock heterogeneous ground. Working with micro-earthquakes in an optics laboratory offers the advantage of using sophisticated electronics and laser optics to study complex seismic effects in small volumes and design systems that can be applied to the macro-world with real earthquakes.

4.3.1 Piezoelectric transducers

The most common technique to generate and detect acoustic waves in solids is by using the reverse and direct piezoelectric effect. Some anisotropic single crystal materials possess the property to get electrically polarized when a mechanical force is applied to them. This is the direct piezoelectric effect. These piezoelectric anisotropic single crystals possess also the property to get deformed when an electrical field is applied to them. This is the reverse piezoelectric effect. We remember that mechanical deformations in solids are described by the strain field tensor with components S_{ij} and are caused by the stress field tensor with components T_{ij}.

This may look as a complicated experiment but in really it is pretty simple to be done in practice. A large class of single crystals that do not have a center of symmetry possesses piezoelectric properties. If a plate cut of such piezoelectric crystal is put between two metal electrodes to which a voltage is applied the plate will deform. Depending on the crystallographic orientation of the plate this deformation can be parallel to the electrical field or perpendicular to it. If the electrical field is an alternative voltage the plate will expand or shrink according to the AC voltage. If the plate is firmly attached to a solid an elastic longitudinal wave will start propagating in that solid. Same thing will happen if the plate deforms in directions perpendicular to the electrical field. A shear wave will be generated in the solid to which this plate is attached. We make the conclusion from the above said that the reverse piezoelectric effect can be used to generate P- and S-waves in solids.

Beside single crystals of noncentrosymmetric classes some ceramics can possess piezoelectric properties. Here is how piezoelectricity works in ceramics. Ceramics are isotropic materials. However, their isotropy is similar to that one of metal. Ceramics like metals have grain structure. This difference between ceramic and metal grains is that metal grains are single crystals formed of metal atoms having cubic symmetry while ceramic grains are formed of metal oxides some with complex chemical composition which are electrical dipoles. Grains of metals do not form dipoles. Many ceramics do not have grain dipoles either. Ceramics which grains are electrical dipoles are not piezoelectric because the dipoles are oriented randomly in the space making the overall ceramic block electrically neutral. If the ceramic is heated at Curie temperature (specific for each ceramic) when the dipoles become mobile and an electrical field is applied all the dipoles get their electrical moments oriented along the direction of the electrical field. The field should be applied during the cooling process until temperature goes under Curie point. The dipole moment will freeze and the ceramic becomes a piezoelectric material.

In the same way the direct effect can be used to detect elastic waves that have been generated in a solid by attaching the same type of piezoelectric plate on the opposite wall of the solid. When the propagating acoustic waves reach this plate they will deform it and the plate will be polarized electrically. If the plate is located between two metal electrodes the electrical signal coming from them can be measured with a voltmeter or better with an oscilloscope on the screen of which we will see the electrical signal displayed in time. We can make a conclusion again that the direct piezoelectric effect can be used to detect P- and S-waves in solids. We note that piezoelectric ceramics can be used only for generation and detecting of P-waves. As will see they can be used for generating S-waves in mode conversion wedges. However, they cannot be used for detecting S-waves. The devices that are used in practice to generate and detect elastic waves are called piezoelectric transducers.

Of course, implementing such a generator or detector of P- or S-waves into practice will require some optimizations to be made. For example if the velocities of propagation of P- and S-waves are known in the direction of the crystal which is perpendicular to the plate then the plate can be cut in a way that its thickness to be equal to half of the acoustic wavelength corresponding to the voltage frequency. This means that the plate will be in mechanical resonance at the frequency of the voltage signal. If the plate operates as a generator it will produce acoustic waves with higher intensity. If the plate is used as a detector its sensitivity will be much higher if it operates under resonance conditions. Another parameter that needs to be taken into account in this experiment is the matching of the elastic impedances of the plate and the solid. They need to have close values if we want to avoid getting reflected acoustic waves.

Piezoelectric transducers for generating and detecting P-waves in solids are available commercially. They have a wide range of applications in non-destructive testing, medical ultrasonics, underwater acoustics, etc. Fig. 4.3.1.1 shows how such a transducer is used to generate P-waves in a medium. These transducers are using piezoelectric ceramic materials (PZT) which chemical composition is a combination of metal oxides such as $PbZrO_3$ or $PbTiO_3$. The composition of the PZT ceramic is formed by single-crystalline micro-domains with a constant dipole momentum. The domains are randomly oriented in

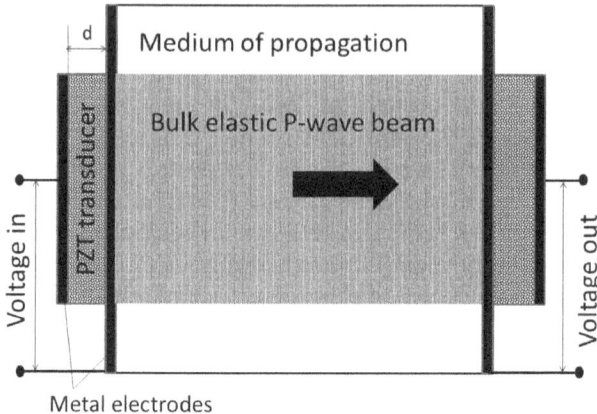

Fig. 4.3.1.1. Ceramic PZT transducers generating and detecting P-waves in a solid.

space so the ceramic is not polarized. At high temperatures above 1100°C if a strong electric field is applied the mobile domains will get oriented along the electric field in same direction; the ceramic is polarized and ready to be used as a transducer. The resonance frequency v_R of the transducer is determined by the thickness d of the PZT applying an AC voltage on the metal electrodes the ceramic transducer expands and contracts in the direction of propagation with the frequency of the electrical field $-v_E$. If the AC signal is tuned at $v_E = v_R = \dfrac{2V_P}{d}$ the system is in resonance and standing waves are formed in the volume of the transducer. The elastic impedance $Z_i = V_i \cdot \rho_i$, where ρ_i is mass density, of the transducer and the medium of propagation should have close values in order to ensure adequate transfer of elastic power from the transducer into the medium of propagation. If $Z_T \gg Z_M$ most of the elastic power of the transducer will be reflected back from the boundary.

If a bulk elastic S-wave has to be generated in the medium or propagation, a different type of transducer has to be used. Piezoelectric effect in non-cubic low-symmetry single crystal such as α-quartz or $LiNbO_3$ from the trigonal crystal symmetry can be used as transducers of S-waves. Both crystals are anisotropic and an AC electric field applied in certain direction is resulting into generation of S-wave. A piezoelectric transducer can be fabricated using such single crystal in a similar way as the ceramic PZT. In this case it is necessary first to orient the crystal using X-ray diffraction and then using a diamond saw to cut a slice from it with appropriate orientation for S-waves. The thickness of the slice has to be determined in the same way as in the case of the PZT to get the desired resonance frequency for S-waves. We note that elastic impedance and velocity of propagation will be different.

Here we will discuss a different approach to generate S-waves that is more appropriate for our micro-earthquake discussion. Let us consider Fig. 4.3.1.2. We notice that our transducer from Fig. 4.3.1.1 is used in this new system formed by a PZT and two different mediums of propagation each with its own elastic impedance.

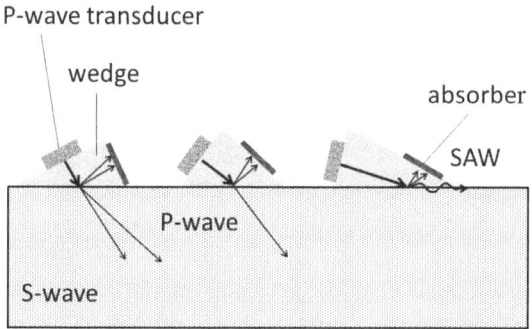

Fig. 4.3.1.2. A wedged PZT transducer generating P-waves, S-waves, and SAW in a solid.

4.3.2 Piezoelectric generation of surface waves

We have discussed the case of SV-wave incident to a free boundary surface under the critic angle. Such wave mode can be converted into surface elastic wave of Rayleigh. A wedge piezoelectric transducer as shown in Fig. 4.3.1.2 can be used to generate Rayleigh surface elastic waves if the wedge is tuned to produce a critical angle of incidence to the free boundary surface. This technique to generate surface acoustic waves is not very efficient because of the low energy output. Using interdigital electrodes on a piezoelectric surface has been proved to be much more efficient way to generate Rayleigh waves because of the possibility to use resonance amplification. Figure 4.3.2.1 shows such an interdigital transducer. Each pair of interdigital electrodes produces a mechanical deformation upon applying electrical voltage to the electrodes because of the reverse piezoelectric effect. The space of a half wavelength between the interdigital electrodes provides the possibility all these mechanical deformations to propagate in phase in the same direction. The resulting surface elastic waves generated by each electrode pair will form a resonance system and an amplified wave will emitted be emitted from the aperture of the transducer.

The generation of a Rayleigh surface acoustic wave was possible using the interdigital transducer in Fig. 4.3.2.1 because of the reverse piezoelectric effect. The reverse piezoelectric effect produces mechanical deformation when electrical field (voltage) is applied to the electrodes. The direct piezoelectric

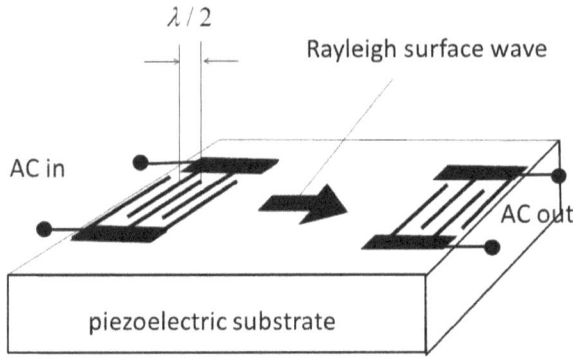

$\lambda / 2$

Rayleigh surface wave

AC in

AC out

piezoelectric substrate

Fig. 4.3.2.1. An interdigital metal electrode generates a Rayleigh wave on a piezoelectric surface.

effect provides the opposite—a voltage is generated if a mechanical deformation is produced. The propagating surface acoustic waves are producing such periodic deformation that if reaching an identical interdigital transducer will generate voltage on the electrodes. If an electrical pulse is sent to the interdigital electrodes in Fig. 4.3.2.1 a pulse surface elastic wave will be generated. An identical interdigital transducer on the other side can detect SAW the same way as we saw with bulk waves. If reflected back by some reflecting boundary this surface elastic wave can be detected by the same interdigital transducer that had generated it. The first and second wave pulses will be separated by a time interval that is proportional to the velocity of propagation of the surface elastic wave and it can be observed on the screen of an oscilloscope.

4.4 Diffraction of light from elastic waves. Laser probe

When an optical beam crosses an acoustic beam a diffraction effect occurs similar to this one from a diffraction grating. The acoustic beam represents a series of more dense and less dense areas moving in the solid that has the same effect on the optical beam as a moving diffraction grating. The optical refractive index of most materials depends linearly of density. The higher the material density, the higher the refractive index is. We note that there is no strict linear relationship between density and optical refractive index. The experiments with various optical glasses however show that some semi-linear correlation between these two parameters exists. At the output of the elastic wave aperture a diffraction pattern can be observed similar to that of a mechanical diffraction aperture Fig. 4.4.1. The incident laser beam gets diffracted by SAW (surface acoustic waves) as well by BAW (bulk acoustic waves). In the case of BAW the laser wavelength has to be selected in such a way that the medium of elastic wave propagation to be transparent. The zero diffraction order corresponds

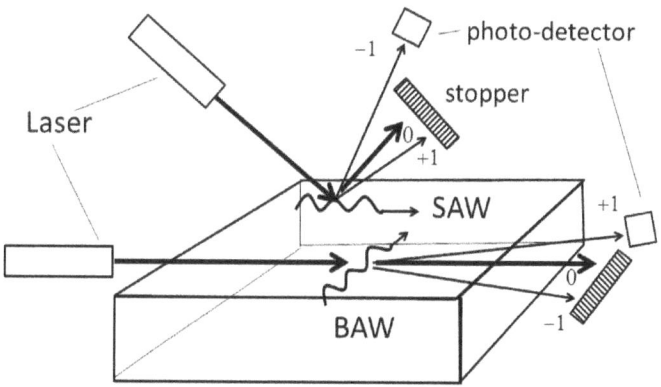

Fig. 4.4.1. Diffraction of a laser beams from BAW and SAW.

to the reflected for SAW and transmitted for BAW beams. The +1 and –1 are the first order diffraction beams. There are higher orders also (not shown in Fig. 4.4.1) - +2, +3,.... as well as –2, –3,... If a screen is installed to collected the diffracted beams diffraction maxima will be seen arranged on a straight line parallel to the direction of propagation for both SAW and BAW. The incident beam, direction of elastic wave propagation, and diffraction maxima all lie in the same plane. Usually a light stopper is used to prevent the zero order laser beam, which is the most intense, from getting close to the photo-detector in order to avoid optical noise. An interesting observation is the dependence of the diffraction angle on the acoustic wavelength and, respectively, on the velocity of propagation. In the case of a diffuse elastic field formed by P- and S-waves propagating in all direction the diffraction maxima will be located on close curves representing the slowness surfaces of all elastic waves. This can be used to identify all bulk elastic waves propagating in the sample.

4.4.1 Optical indicatrices

Figure 1.3.2 shows the optical axes and optical indicatrices (surfaces of optical indexes) of isotropic and anisotropic materials. The general equation of the optical indicatrix is an ellipsoid which main axes are the three optical indices:

$$\frac{x_1^2}{n_1^2} + \frac{x_2^2}{n_2^2} + \frac{x_3^2}{n_3^2} = 1 \qquad (4.4.1.1)$$

The ellipsoid has to circular sections. The axes that are perpendicular to these circular sections are called optical axes. The crystal shown in Fig. 1.3.2 has one optical axis and the ellipsoid's equation is:

$$\frac{x_1^2 + x_2^2}{n_O^2} + \frac{x_3^2}{n_E^2} = 1 \tag{4.4.1.2}$$

n_O is the optical index along the axes x_1 and x_2 called ordinary index, while n_E is the optical index along the x_3-axis called extraordinary index. The isotropic case has in all direction the same refraction index.

The general formula of an indicatrix can be presented as:

$$B_{ij} x_i x_j = 1$$

If the crystal is strained by some external stress the indicatrix will be deformed:

$$\Delta B_{ij} = p_{ijkl} S_{kl} \tag{4.4.1.3}$$

The dimensionless coefficients p_{ijkl} form a tensor of rank 4 called elasto-optical tensor. S_{kl} is the strain tensor.

ΔB_{ij} can be presented also as function of the stress tensor T_{kl}:

$$\Delta B_{ij} = \pi_{ijkl} T_{kl} \tag{4.4.1.4}$$

The tensor π_{ijkl} is called piezo-optical tensor.

Since $T_{kl} = c_{klmn} S_{mn}$ we get:

$$p_{ijmn} = \pi_{ijkl} c_{klmn} \tag{4.4.1.5}$$

ΔB_{ij} can be presented also using the dielectric permittivity constant ε_{il} because $n = \sqrt{\varepsilon_r \mu_r}$, where $\varepsilon_r = \varepsilon/\varepsilon_0$ and $\mu_r = \mu/\mu_0$ are the dielectric and magnetic permittivity constants. For a nonmagnetic material $\mu_r = 1$ and, therefore, $n = \sqrt{\varepsilon_r}$ or

$$\varepsilon_{ij} B_{jk} = \delta_{ik} \tag{4.4.1.6}$$

Differentiating Eq. 4.4.1.6 yields to:

$$\Delta \varepsilon_{ij} B_{jk} + \varepsilon_{ij} \Delta B_{jk} = 0$$

or

$$\Delta \varepsilon_{il} = -\varepsilon_{ij} p_{jkmn} \varepsilon_{kl} S_{mn} \tag{4.4.1.7}$$

Since the dielectric tensor (and the optical index tensor) are functions of the strain elastic wave that propagates through the material will cause a periodical changes of the indicatrix. If a light beam goes though that material and crosses the elastic beam an elasto-optic interaction will occur. Depending on the incidence angle there are two possible cases: 1) Raman-Nath diffraction for an incidence normal and 2) Bragg diffraction for an incidence under Bragg's angle.

4.4.2 Raman-Nath diffraction

The geometry of the case of a normal incidence is presented in Fig. 4.4.2.1. The frequencies of diffracted light at various diffraction orders are sums or differences from the frequencies of the incident light and the elastic wave. The diffraction angle is defined by

$$\sin \theta_m = \pm m \frac{\lambda_L}{\lambda_{EW}} \qquad (4.4.2.1)$$

where $m = \pm 1, \pm 2, \pm 3,....$ the diffraction orders.

The intensity of the light in various diffraction orders follows Bessel functions as shown in Fig. 4.4.2.2.

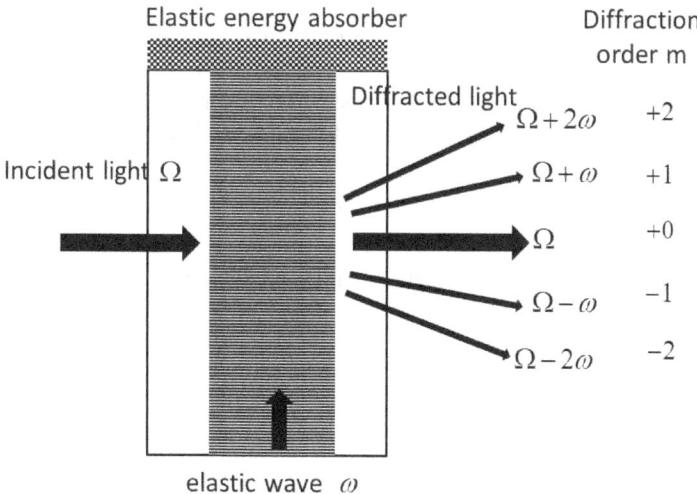

Fig. 4.4.2.1. Raman-Nath diffraction at normal incidence.

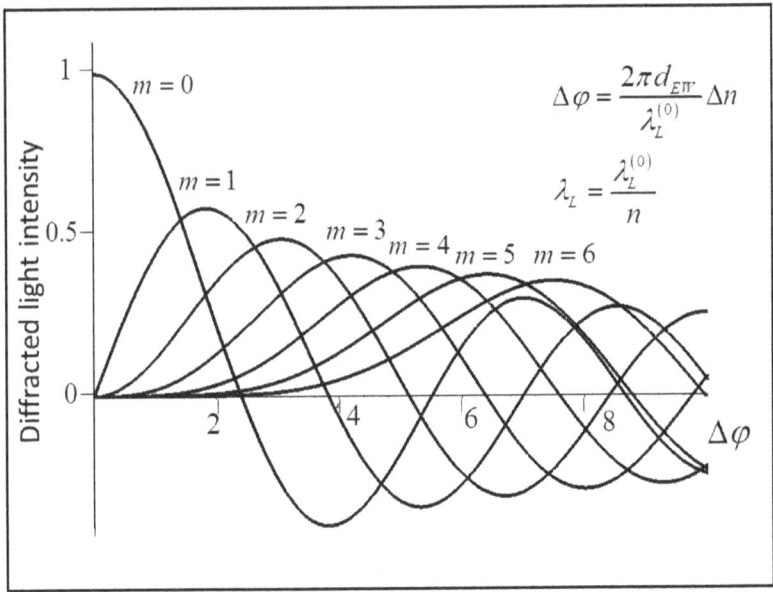

Fig. 4.4.2.2. Intensity of the diffraction orders in Raman-Nath elasto-optic interaction.

4.4.3 Bragg diffraction

The Bragg diffraction creates only one diffracted maximum. The angle of incidence θ_B allows constructive interference of the first diffraction order only as shown in Fig. 4.4.3.1. The first order beams are in-phase and interfere constructively if the angle of incident is equal to Bragg's angle defined by:

$$\sin \theta_B = \frac{\lambda_L}{2\lambda_{EW}} \qquad (4.4.3.1)$$

For an isotropic material the elasto-optic tensor is:

$$
\begin{pmatrix}
p_{11} & p_{12} & p_{12} & 0 & 0 & 0 \\
p_{12} & p_{11} & p_{12} & 0 & 0 & 0 \\
p_{12} & p_{12} & p_{11} & 0 & 0 & 0 \\
0 & 0 & 0 & \dfrac{p_{11}-p_{12}}{2} & 0 & 0 \\
0 & 0 & 0 & 0 & \dfrac{p_{11}-p_{12}}{2} & 0 \\
0 & 0 & 0 & 0 & 0 & \dfrac{p_{11}-p_{12}}{2}
\end{pmatrix}
\qquad (4.4.3.2)
$$

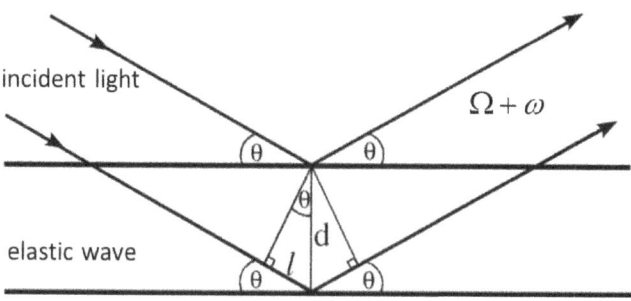

Fig. 4.4.3.1. Bragg Diffraction in Elasto-Optic Interaction.

From Eq. 4.3.1.7 for a strain S_{11} (a P-wave propagating along the x-axis) in an isotropic material $\Delta\varepsilon_{il} = -\varepsilon^2 p_{ilmn} S_{mn} = --\varepsilon^2 p_{il11} S_{11}$ the only elasto-optic coefficient $p_{il11} \neq 0$ are $p_{1111} = p_{11}$, $p_{2211} = p_{21}$, $p_{3311} = p_{31}$. Therefore we have $\Delta\varepsilon_{11} = -\varepsilon^2 p_{11} S_{11}$ and $\Delta\varepsilon_{22} = \Delta\varepsilon_{33} = -\varepsilon^2 p_{12} S_{11}$ with $\Delta\varepsilon_{il} = 0$ if $i \neq l$. The refractive indices are respectively equal to $n_1 = \sqrt{\varepsilon + \Delta\varepsilon_{11}}$ and $n_2 = n_3 = \sqrt{\varepsilon + \Delta\varepsilon_{22}}$. Since $\varepsilon \gg \Delta\varepsilon_{il}$ we can develop the refractive index in Taylor series:

$$n_i \simeq n\left(1 + \frac{\Delta\varepsilon_{il}}{2\varepsilon}\right) = n + \Delta n_i \text{ with } \Delta n_i = \frac{\Delta\varepsilon_{il}}{2n}$$

Therefore we have:

$$\Delta n_1 = -\frac{n^3}{2} p_{11} S_{11} \qquad (4.4.3.3)$$

$$\Delta n_2 = \Delta n_3 = -\frac{n^3}{2} p_{12} S_{11} \qquad (4.4.3.4)$$

In the case of an S-wave propagating along the x-axis with a polarization along the z-axis the strain is S_{13}.

$$\Delta\varepsilon_{il} = -\varepsilon_{ij} p_{jkmn} \varepsilon_{kl} S_{mn} = -\varepsilon_{ii}\varepsilon_{ll} p_{il13} S_{13}$$

The only non-zero component of the elasto-optic tensor is $p_{1313} = p_{55}$ and the tensor $\Delta\varepsilon_{il}$ is:

$$\Delta\varepsilon_{il} = \begin{pmatrix} 0 & 0 & \varepsilon_{13} \\ 0 & 0 & 0 \\ \varepsilon_{31} & 0 & 0 \end{pmatrix}$$

$$\Delta\varepsilon_{13} = -\varepsilon_{11}\varepsilon_{33} p_{55} S_5$$

The power density of the elastic wave is (Dieulesaint and Royer 1974):

$$P_{EW} = \frac{1}{2} \rho V_{EW}^3 S^2 \qquad (4.4.3.5)$$

From $\Delta \varphi = \frac{2\pi d_{EW}}{\lambda_L^{(0)}} |\Delta n|$ and Eq. (4.4.3.5) we get:

$$|\Delta n| = \frac{n^3}{2} pS = \sqrt{\frac{p^2 n^6}{2 \rho V^3} P} = \sqrt{\frac{MP}{2}} \qquad (4.4.3.6)$$

In Eq. 4.4.3.5 M is called merit factor of the material.

The intensity of the diffracted light from the elastic wave is:

$$I = I_0 \sin^2 \left(\pi \frac{d_{EW}}{\lambda_L^{(0)}} \sqrt{\frac{MP}{2}} \right) \qquad (4.4.3.7)$$

Equation 4.4.3.7 shows that the higher the merit factor of the material the lower is the elastic power necessary to deflect the optical beam.

4.5 Laser generation of elastic waves

Using piezoelectric transducer in contact with the medium of propagation is an effective technique to generate bulk shear and longitudinal waves in a narrow frequency bands. Seismic waves that propagate in heterogeneous media could have wide frequency spectrum. For better modelling systems for reflection of elastic waves it would be more appropriate to use wide spectrum transducers or operate the piezoelectric transducers in a pulse mode. Instead of sending to the electrodes a continuous signal with a frequency that is close to the resonance frequency of the transducer the electrical signal's amplitude could be modulates in form of pulses full of high frequency signal. There are also other optical techniques that can be used to generate elastic wave pulses in solids imitating seismic wave pulses in a better way.

If a laser pulse reaches the surface of a solid that absorbs light with the laser's wavelength the absorbed optical energy will create a pulse of heat in the point of absorption. The resulting thermal expansion will cause a local deformation and an elastic wave pulse will propagate through the solid. Thermo-elastic effect is based on expansion of elastic materials when heat is absorbed. The most efficient way to transfer heat to a material substrate is by using a focused laser beam that gets absorbed on the surface raising locally temperature and thus causing thermal expansion (Fig. 4.5.1). If the laser beam's intensity is modulated the thermal expansion will be also modulated resulting in pulse waves propagation through the material. Any laser which wavelength is absorbed by the sample can be used. If the material does not absorb well

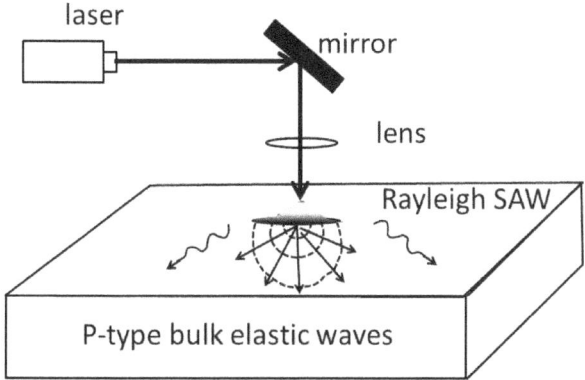

Fig. 4.5.1. Experimental setup for generation of elastic pulses using photothermal effect.

the laser beam at that wavelength an absorbing material can be coated on the sample surface such as ink. Using pulsed lasers such as solid state infrared Q-switched Nd:YAG (Neodymium: Yttrium Aluminum Garnet Nd:$Y_3Al_5O_{12}$) or gas lasers—Excimers (UV), CO_2 lasers is an efficient way to generate elastic pulses in a solid. High intensity lasers can produce significant thermo-elastic effect or even ablation (evaporation of the substrate by the heat pulse).

Bulk longitudinal elastic waves can be generated in a solid as well as Rayleigh surface elastic waves. Bulk waves have spherical fronts and can be generated from a point source. Rayleigh surface waves can be generated more efficiently from a line thermo-elastic source. In this case instead of spherical optical lens a cylindrical lens can provide a thin laser line source. If the surface of the solid sample is covered with a film with elastic properties matching the conditions of waveguide propagation the laser can generate Love waves.

4.5.1 Laser-generated solitary surface pulses

Pulsed laser beam is the most efficient way to generate powerful surface elastic pulses. This technique has been used widely in the study of the nonlinear properties of surface elastic waves and the waveform evolution during the wave propagation. Elastic solitary surface pulses cannot be excited in nondispersive media of propagation. With a finite-amplitude elastic wave we can get into a nonlinear regime of propagation in a nondispersive solid with evolution of the waveform, but we cannot get a solitary pulse. We have seen that Raleigh waves get dispersive if the surface of the substrate of propagation is coated with a layer with different elastic properties. The level of dispersion can be controlled by selection materials with appropriate elastic impedances.

To create a solitary pulse the amplitude of the elastic strain field should be high enough. For example to obtain nonlinear regime of propagation with uncoated nondispersive material the Mach number should be of the order of $M \approx 3.10^{-3}$ (Lomonosov and Hess 1999). At higher amplitudes in nondispersive substrates shock waves cause cracks. $M \approx 2.10^{-2}$ is the upper strength limit for most materials. The dispersion suppresses the formation of shock waves, so the crack threshold is higher in a dispersive substrate coated with a layer. (Lomonosov et al. 2002).

The experimental setup for exciting solitary elastic surface pulses is shown in Fig. 4.5.1.1 (Lomonosov et al. 2002). A Q-switch Nd:YAG 8nsec laser pulsed beam with wavelength 1.064 μm is focused through a cylindrical lens to a 7 mm long and 30 μm wide line. The laser pulse energy is between 30–60 mJ. To increase the efficiency of the photoelastic process a light absorbing carbon film is applied in the laser focus area (Lomonosov and Hess 1999; Lomonosov et al. 1999). The resulting thermoelastic shock launches a powerful surface elastic pulse. The propagated pulse is detected using an optical probe (Section 4.4). The output signal of the position-sensitive detector is proportional to the surface slope $\dfrac{\partial u_3}{\partial x_1} = -\dfrac{1}{V_R}\dfrac{\partial u_3}{\partial t}$ in the direction of propagation x_1 (Fig. 4.5.1.1) with u_3 being the vertical displacement along the x_3-axis and V_R - the Raleigh velocity. The typical characteristic of 'Mexican hat' is observed for surface pulse propagation on the Si (100)-plane in the direction <100>. The pulse propagation on Si (111)-plane in the <112>-direction the shape of the pulse is very different from the case of Si (100)-plane with direction of propagation <100>. The difference between the two geometries demonstrates the strong influence of the anisotropy of the substrate on the shape of the solitary surface pulses (Lomonosov et al. 2002).

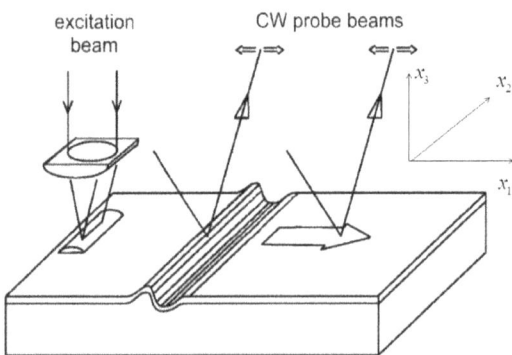

Fig. 4.5.1.1. Experimental setup for pulsed laser excitation of solitary pulses (Lomonosov et al. 2002).

4.6 Experimental modelling of SAW deflectors

If the medium of propagation shown in Fig. 4.5.1 has a heterogeneous structure, the laser probe can be used in various points of the sample to identify the type and intensity of elastic waves (Fig. 4.6.1).

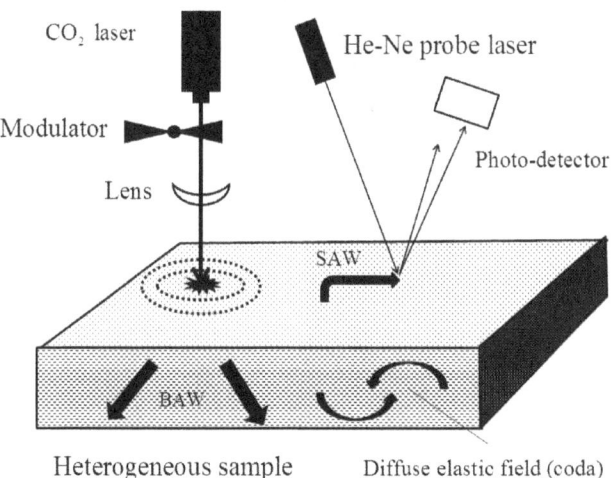

Fig. 4.6.1. Laser generation of simulated mini-earthquake in a heterogeneous sample.

Figure 4.6.2 presents an experimental setup to evaluate the efficiency of an earthquake defensive system for Love waves. Love waves are generated in the thin layer on top of the basic substrate if the HS-wave in the layer has slower velocity of propagation than the HS-wave in the substrate. At a certain point the waveguide structure is interrupted by a set of groves. We want to compare the elastic amplitudes of the Love waves before they reach the interrupting groves of the shield and after the shield. A shield stopping Love waves can be a simple interruption of the waveguide layer—for example a grove dug into the medium of propagation with a depth and width of the order of the Love wave wavelength and filled with some metamaterial having a different elastic impedance.

Figure 4.6.3 presents similar experimental setup as Fig. 4.6.2 that can be used for the design of a shield for deflecting Raleigh surface acoustic waves. The shield can be designed to reflect back Rayleigh waves or convert them into body waves directed down the Earth's crust. The efficiency of the shield can be determined experimentally by using a signal comparator providing data about the intensity of the Rayleigh waves before and after the shield.

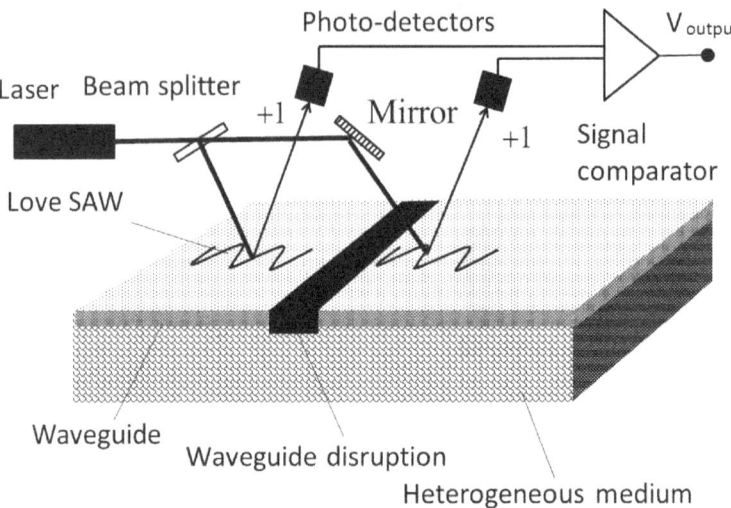

Fig. 4.6.2. Experimental setup to evaluate the efficiency of an earthquake defensive system for Love waves.

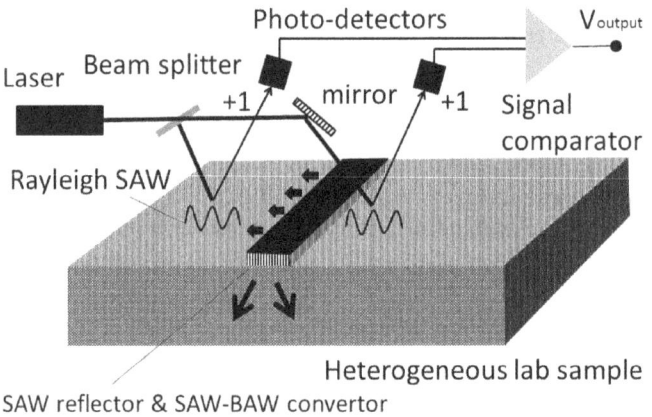

Fig. 4.6.3. Experimental setup to measure reflected Rayleigh waves in a heterogeneous sample.

The experimental data obtained with help of the acoustooptics experiments shown in Figs. 4.6.2 and 4.6.3 can be used as an input to computer simulation models optimizing the design of the shield as well as testing the efficient of metamaterials used.

Similar experimental modelling setups with laser generation of elastic pulses and laser probe detection can be constructed using surface-to-body mode conversion groves and scattering periodic structures shown in Figs. 3.5.1, 3.6.2.2, 3.6.2.3, and 3.6.2.4. Since the calculations are very complex in the case of nonlinear SAW propagation on heterogeneous, dispersive, and dissipative substrates such experimental modelling techniques can lead to useful practical applications and shield construction designs. As shown in Fig. 3.5.1 a simple concrete step with a height equal to half of Rayleigh wavelength built around a city can reduce the amplitude of upcoming Rayleigh waves by 75%.

A schematic presentation of the idea to shield a city from upcoming seismic waves is shown in Fig. 4.6.4. A great number of design solutions exist for reflecting back, absorbing, or redirecting down into the crust upcoming body or surface seismic waves. Various metamaterials can be developed to achieve these goals. The simplest solutions are to look for appropriate reflecting structures and metamaterials with appropriate elastic impedances, but also more sophisticated designs could include development of nonlinear metamaterials able to suppress nonlinear parameters in the seismic wave propagation that contribute to earthquakes.

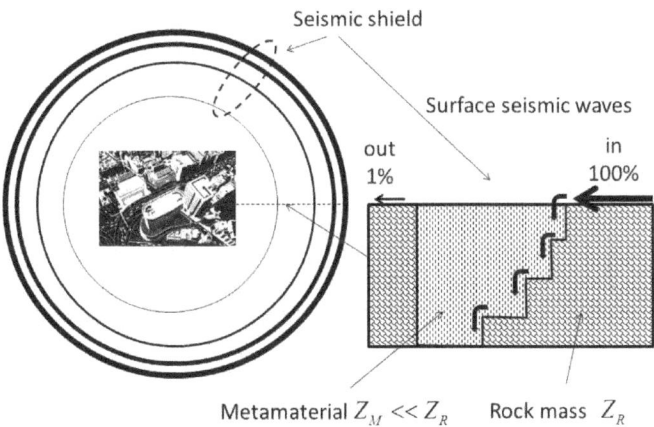

Fig. 4.6.4. Seismic shield.

Conclusion

Seismic wave are highly nonlinear elastic waves that propagate in a highly nonlinear, heterogeneous, and dispersive media. Using the tools of the classical seismology—continuum mechanics and general scattering theory—leave many questions about seismic wave propagation unanswered because these theories are linear and neglect parameters of second, third, and higher orders. Rock is a highly nonlinear medium where quadratic and cubic terms often are dominant factor in shaping elastic properties of rocks. Seismic waves are high-amplitude elastic waves that change their waveform as they propagate, so linear Fourier analysis is not suitable to describe their behavior. During their propagation they cause uniaxial deformation of the rock making the rock to behave as an anisotropic medium. Nonlinearity of elastic wave propagation cause the growth of higher harmonics, subharmonics, and waves of combination frequency that are critical for self-modulation, self-focusing, parametric amplification, soliton generation, wave-from-wave scattering and other phenomena that contribute to the seismic wave propagation. These phenomena are specific to nonlinear wave propagation and are left undetected by the linear theories. The generation and propagation of coda waves cannot be explained without taking into consideration nonlinear phenomena such as combination frequency wave generation and self-modulation. If a seismic wave were decomposable in an unlimited number of high harmonics that interact with each other, according to the linear theory that would cause an unlimited growth of harmonics that interact with each other which will end to a shock wave, which never happens in reality because of the nonlinearity of the propagation that suppresses the interaction among the harmonics. Earthquakes with characteristics as we know them cannot occur in a linear world. This means that they have to be analyzed using the tools of the nonlinear elastic theory of elastic wave propagation. Rock is a highly nonlinear medium also that affect the finite-amplitude elastic wave propagation. Nonlinearity phenomena due to cracks full of fluid, random heterogeneities, cavities, pockets of high strain storing significant amounts of potential energy that can be released into a passing seismic wave increasing its energy density, and soil basins able to create local resonance conditions contribute significantly to the dispersion, dissipation, trapping, scattering, or amplification of seismic waves. The problem is that the nonlinear wave propagation theory is still under development and many nonlinear phenomena have not been fully understood. However, advanced computer simulation, experimental modelling complimentary to acoustoelectronics and acoustooptics experimental modelling techniques using laser generation and detection of elastic waves, as well as the development of metamaterials can accelerate the progress of the development of earthquake-resistant design and technology. From many earthquakes in the past it has become clear that just building reinforced man-made constructions is not enough to fight destructive seismic forces. Instead of facing the overwhelming power of seismic waves developing shield systems around cities and towns located in frequent-earthquake areas could prove to be an efficient strategy to prevent loss of life and destruction. Shield constructions able to redirect or absorb seismic body and specifically surface waves can be developed using combination of computer simulation, experimental design, and metamaterials. Metamaterials designed with specific nonlinear and dispersion characteristics can be inserted in some geological structures around cities with known seismic resonance or amplification parameters to prevent strengthening of crossing seismic waves. Since body seismic waves rarely affect directly man-made constructions, most of the attention should be put on the mechanisms of generation of surface elastic waves by body waves as well as the geological conditions enabling mode conversion and energy transfer. Many problems of SAW generation and propagation as well as body-to-surface and surface-to-body wave conversions have been solved in acoustoelectronics and can be successfully implemented in practical seismic solutions, however many others need to be addressed. An example are Love seismic waves. These waves with horizontal polarization in the plane of propagation are notoriously destructive to man-made constructions. However, Love waves are dispersive waves that require specific geometric structures forming elastic waveguides to exist. Natural waveguides are rare in the Earths geological systems. No other surface waves with horizontal polarization have been considered in the linear seismology theory. However, nonlinearity and material dispersion allow the generation of surface elastic waves with Love-like polarization which do not require any geometrical dispersion structures such as layered elastic waveguide to exist —skimming waves. These waves can be generated anywhere on the Earth's surface similarly to Rayleigh waves with much higher probability than Love waves and carry out destructions to man-made constructions similar to Love waves. Another example are solitary seismic waves and seismic solitons. These types of body and surface waves can exist only in conditions of nonlinearity and dispersion. Water tsunami waves which have soliton characteristics are notorious with their distinctiveness. Seismic solitons has not been proven to exist, but mathematical solutions to the governing equations of waveform evolution show that solitons can exist on solid surfaces in conditions of high nonlinearity and dispersion, are stable, and can travel long distances surviving collisions with each other. Taking into account the complexity of the problem it will take a while until the propagation of nonlinear dispersive seismic waves are fully understood. Meanwhile simple efficient seismic-wave shielding constructions can be designed using experimental modelling and offered to city planning organization, developers, and civil engineering companies for implementation around cities with frequent earthquake activity.

During the past decade nonlinear wave propagation and nonlinear theory of elasticity have been in process of intense research in various scientific fields such as medical imaging, optoelectronics, electrical engineering, and modern seismology. Some problems have been solved, but many others remain unsolved, so the research on nonlinear phenomena will grow exponentially. Earthquake-resistant design and technology can be developed only if nonlinear seismic wave propagation phenomena are taken into consideration despite the thorny road toward solving nonlinear governing equations. Linear theories are simple and elegant, however our world is nonlinear and so are the seismic waves causing earthquakes.

References

Aaronson, S.L., T. Barkve, J. Tjotta and S. Tjotta. 1984. Distortion and harmonic generation in the nearfield at finite-amplitude sound beam. J. Acoust. Soc. Am., 75: 744–768.

Akhmanov, S.A. and R.V. Khokhlov. 1964. Problems of nonlinear optics, Viniti, Moscow.

Aki, K. 1972. Earthquake mechanism, Tectonophysics, Elsevier B.V., 13: 423–446.

Aki, K. and B. Chouet. 1975. Origin of coda waves: Source, attenuation, and scattering effects, Journ. Geophys. Res., 80: 3322–3342.

Aki, K. and P.G. Richards. 1980a. Quantitative seismology: Theory and methods, W.H. Freeman & Co., San Francisco.

Aki, K. and P.G. Richards. 2002b. Quantitative seismology, Sausalito, California: University Science Books.

Albregtsen, F. 2008. University of Oslo, Reflection, refraction, diffraction, and scattering.

Andreev, N.N. 1955. On some values of the second order in acoustics. Acust. Zh., 1: 3–11.

Andreev, V.G., O.A. Vasilieva, E.A. Lapshin and O.V. Rudenko. 1985. Akust. Zh., 31: 12–18.

Auld, B.A. 1973. Acoustic fields and waves in solids, John Wiley & Sons, New York.

Barford, L.A., R.S. Fazzio and D.R. Smith. 1992. An introduction to wavelets, Hewlett Packard, HPL: 92–124.

Beer, F., R. Johnston, J. Dewolf and D. Mazurek. 2009. Mechanics of materials, New York: McGraw-Hill companies.

Ben-Menahem, A. and S.J. Singh. 1981. Seismic waves and sources, Springer-Verlag.

Beresnev, I.A. and V.N. Nikolaevskiy. 1993. A model for nonlinear seismic waves in a medium with instability, Physica D: Nonlinear Phenomena, 66: 1–6.

Biryukov, S.V., Y.V. Gulyaev, V. Krylov and V. Plessky. 1995. Surface acoustic waves in inhomogeneous media, Series: Springer Series on Wave Phenomena, vol. 20.

Brady, A.G., V. Prez and P.N. Mork. 1980. Digitization and processing of main-shock ground-motion data from the U.S. geological survey accelerograph network. The Imperial Valley, CA, earthquake of October 15, 1979, 1254: 385–406.

Brûlé, S., E.H. Javelaud, S. Enoch and S. Guenneau. 2014. Experiments on seismic metamaterials: molding surface waves, Physical Review Letters, 112: 1339011–1339015.

Brun, M., S. Guenneau and A.B. Movchan. 2009. Achieving control of in-plane elastic waves, Appl. Phys. Lett., 94: 061901–061903.

Brugger, K. 1964. Thermodynamic definition of higher order elastic coefficients. Phys. Rev., A133: 1611–1612.

Byers, W. G. 1976. China Collection, Pacific Earthquake Research Center, University of California, Berkeley.

Cedroits, A.A. and V.A. Krasil'nikov. 1963. Finite-amplitude elastic waves in solids and deviation from Hooke's law. Sov. Phys. JETP, 16: 1122–1126.

Clough, G.W., J.R. Martin, II and J.L. Chameau. 1994. The geotechnical aspects, Practical lessons from the Loma Prieta earthquake, National Academies Press, 29–32.

Chakraborty, A. and D. Okaya. 1995. Frequency-time decomposition of seismic data using wavelet-based methods, Geophysics, 60: 1906–1916.

Chapman, C. 2004. Fundamentals of seismic wave propagation, Cambridge University Press, Cambridge.

Curie, J. and P. Curie. 1880. Développement par compression de l'électricité polaire dans les cristaux hémièdres à faces inclinées (Development, via compression, of electric polarization in hemihedral crystals with inclined faces), Bulletin de la Société minérologique de France, 3: 90–93.

De, Krishna. 1970. Finite Strain Theory of Love Waves, Pure Appl. Geophys, 80: 114–123.

Dieulesaint, E. and D. Royer. 1974. Ondes Elatiques dans les Solides, Masson et Cie, Editeurs, Paris.

Dimitrienko, Y. 2011. Nonlinear Continuum Mechanics and Large Inelastic Deformations, Springer.

Eckl, C., A.P. Mayer and A.S. Kovalev. 1998. Do surface acoustic solitons exist? Phys. Rev. Lett., 81: 983–986.

Eckl, C., A.S. Kovalev, A.P. Mayer, A.M. Lomonosov and P. Hess. 2004. Solitary surface acoustic waves, Phys. Rev. E 70: 0466041-04660415.

Ewing, W.M., W.S. Jardetzky and F. Press. 1957. Elastic waves in layered media, Mc Graw-Hill Book Company, New York.

Farhat, M., S. Guenneau and S. Enoch. 2009. Ultrabroadband elastic cloaking in thin plates. Phys. Rev. Lett., 103: 024301–024304.

Farhat, M., S. Guenneau, S. Enoch and A.B. Movchan. 2009. Cloaking Bending Waves Propagating in Thin Elastic Plates. Physical Review B, 79: 0331021–0331024.

Farnell, G.W. 1970. Properties of surface elastic waves. In: W.P. Mason (ed.). Physical Acoustics. Academic Press, New York., 6: 109–166.

Farnell, G.W. and E.L. Adler. 1972. Elastic wave propagation in thin layers. pp. 35–127. In: W.P. Mason and R.N. Thurston (eds.). Physical Acoustics, Vol. IX. Academic Press, New York.

Farnell, G.W. 1978. Type and properties of surface waves. pp. 13–59. In: A.A. Oliner (ed.). Acoustic Surface Waves. Springer-Verlag.

Fjær, E., R.M. Holt, P. Horsrud, A.M. Raaen and R. Risnes. 1992. Petroleum related rock mechanics: Elsevier Science Publishing Co., Inc.

Frenzel, L.E. 2008. Principles of electronic communication systems, McGraw-Hill.

Gilbert, F. and L. Knopoff. 1960. J. Geophys. Res., 65: 3437–44.

Godin, O.A. 2013. Rayleigh scattering of a spherical sound wave. J. Acoust. Soc. Am. 133: 709–720.

Gol'dberg, Z.A. 1960a. On self-interaction of plane longitudinal and transversewaves. Sov. Phys. Acoust., 6: 307–310.

Gol'dberg, Z.A. 1961b. Interaction of plane longitudinal and transverse elastic waves. Sov. Phys. Acoust. 6: 306–310.

Grady, D. 1997. Physics and modeling of shock-wave dispersion in heterogeneous composites. J. Phys. France IV, Supplement au Journal de Physique III, C3: 669–674.

Grady, D. 1998. Scattering as a mechanism for structured shock waves in metals. J. Mech. Phys. Solids, 46: 2017–2032.

Gusev, V.E., W. Lauriks and J. Thoen. 1998. New evolution equations for the nonlinear surface acoustic waves on an elastic solid of general anisotropy. J. Acoust. Soc. Am., 103: 3203–3215.

Gusev, V.E., W. Lauriks and J. Thoen. 1997. Theory for the time evolution of nonlinear Rayleigh waves in an isotropic solid. Phys. Rev., B55: 9344–9347.

Kim, S.H. and M.P. Das. 2012. Seismic waveguide of methamaterials. Mod. Phys. Lett., B26: 12501051–12501058.

Kumon, R.V., M.F. Hamilton, Y.A. Il'inskii and E.A. Zabolotskaya. 1998. Pulsed nonlinear surface acoustic waves in crystals, 16th International Congress on Acoustics, 135 Meeting of the Acoustical Society of America, Seattle, WA, June 20–26, 1998, paper 3aPAb6.

Hamilton, M.F. 1986. Fundamentals and applications of nonlinear acoustics. pp. 1–28. *In*: T.W. Wright (ed.). Nonlinear Wave Propagation in Mechanics. Am. Soc. of Mech. Eng., New York.

Hamilton, M.F., Y.A. Il'inskii and E. Zabolotskaya. 1999. Nonlinear surface acoustic waves in crystals. J. Acoust. Soc. Am., 105(2) Pt.1 (1999), 639–651.

Hao, H.Y. and H.J. Maris. 2001. Experiments with acoustic solitons in crystalline solids. Phys. Rev. B, 64: 0643021–0643027.

Haran, M.E. and B.D. Cook. 1983. Distortion of finite-amplitude ultrasound in lossy media. J. Acosut. Soc. Am. 73: 774–777.

Hokstad, K. 2004. Nonlinear and dispersive acoustic wave propagation. Geophysics, 69: 840–848.

Hudson, J.A. 1977. Scattered waves in the coda of P. J. Geophys., 43: 359–374.

Hudson, J.A. 1980. The Excitation and Propagation of Elastic Waves. Cambridge Univ. Press, Cambridge, UK.

Hudson, J.A. and J.R. Heritage. 1982. The use of the Born approximation in seismic scattering problems. Geophys., J. R. Astron. Soc., 66: 221–240.

Hudson, J.A., R.F. Humphryes, I.M. Mason and V.K. Kembhavi. 1973. J. Phys. D: Appl. Phys., 6: 2174–86.

Johnson, P.A. and T.J. Shankland. 1989. Nonlinear generation of elastic waves in granite and sandstone: Continuous wave and travel time observations. J. Geophys. Res., 94: 17729–17733.

Johnson, P.A. and P.N.J. Rasolofosaon. 1996. Nonlinear elasticity and stress-induced anisotropy in rock. Journal of Geophysical Research, 101: 3113–3124.

Kanamori, H. 1978. Quantification of earthquakes. Nature, 271: 411–414.

Kawahara, Y., H. Ito, M. Shinohara and H. Kawakatsu. 1990. Small-array observation of seismic coda waves in Izo-Ohshima. Zisin (in Japanese). 43: 359–371.

Kawahara, Y., H. Ito, T. Ohminato, H. Kawakatsu, T. Kiguchi and T. Miyazaki. 1991. Analysis of short period seismic coda waves observed with a small-span array (in Japanese). Prog. Abs. Seism. Soc. Japan., 2: 244–256.

Kolomenskii, A.A., A.M. Lomonosov, R. Kuschnereit, P. Hess and V.E. Gusev. 1997. Laser generation and detection of strongly nonlinear elastic surface pulses. Phys. Rev. Lett., 79: 1325–1328.

Kolomenskii, A.A., V.A. Lioubimov, S.N. Jerebtsov and H.A. Schuessler. 2003. Nonlinear surface acoustic wave pulses in solids: Laser excitation, propagation, interactions (invited). Review of Scientific Instruments, 74: 448–452.

Kosevich, Yu. A. 1990. Nonlinear shear surface waves at interfaces and planar defects of crystals, Phys. Lett., A146: 529–534.

Krylov, V.V. 1993. On nonlinear parametric amplification of Rayleigh waves. Phys. Lett. A, 173: 209–2013.

Lamb, H. 1904. The propagation of tremors over the surface of an elastic solid. Phil. Trans. R. Soc., A, vol. 203, London.

Landau, L.D. and E.M. Lifshitz. 1959. Theory of Elasticity, Pergamon Press, New York.

Landau, L.D. and E.M. Lifshitz. 1986. Hydrodynamics, Moscow, Nauka.

Landau, L.D. and E.M. Lifshitz. 1987. Theory of Elasticity, Moscow, Nauka.

Lardner, R.W. 1984. Nonlinear Rayleigh waves: Harmonic generation, parametric amplification, and thermoviscous damping. J. Appl. Phys., 55: 3251–3260.

Leibfried, G. and W. Ludwig 1960. Z. Phys., 80: 160–167.

Love, A.E.H. Some problems of geodynamics 1911, first published in 1911 by the Cambridge University Press and published again in 1967 by Dover, New York, USA (Chapter 11: Theory of the propagation of seismic waves).

Lomonosov, A.M. and P. Hess. 1999. Effects of nonlinear elastic surface pulses in anisotropic silicon crystals. Phys. Rev. Lett., 83: 3876–3879.

Lomonosov, A., V.G. Mikhalevich, P. Hess, E. Yu. Knight, M.F. Hamilton and E.A. Zabolotskaya. 1999. Laser-generated nonlinear Rayleigh waves with shocks. J. Acoust. Soc. Am. 105: 2093–2096.

Lomonosov, A.M., P. Hess and A.P. Mayer. 2002. Observation of solitary elastic surface pulses, Phys. Rev. Lett., 88: 0761041–0761044.

Lu, C. 1974. Mass determination with piezoelectric quartz crystal resonators. L. Vac. Sci. Technol., 12: 578–583.

Malvern, L.E. 1969. Introduction to the Mechanics of a Continuous Medium. Prentice-Hall, Engelwood Cliffs, NJ.

Maradudin, A.A. and A.P. Mayer. 1990. Nonlinear waves in solid state physics. A.D. Boardman,T. Twardowski and M. Bertolotti (eds.). (Plenum, New York), p. 113.

Mase, G.E. 1970. Continuum Mechanics, McGraw-Hill Professional.

Mayer, A.P. 1991. Evolution equation for nonlinear Bleustein-Gulyaev waves. Int. J. Eng. Sci., 29: 999–1004.

Mayer, A.P. 1995. Surface Acoustic Waves in Nonlinear Elastic Media, Physics Reports, 256: 237–366, Elsevier.

Mayer, A.P. 2008. Nonlinear surface acoustic waves: Theory, Ultrasonics, 48: 478–481.

McCall, K.R. and R.A. Guyer. 1996. A new theoretical paradigm to describe hysteresis, discrete memory, and nonlinear elastic wave propagation in rock, Nonlinear Processes in Geophysics, 3: 89–101.

Meegan, G.D., P.A. Johnson, R.A. Guyer and K.R. McCall. 1993. Observations of non-linear elastic wave behavior in sandstone. J. Acoust. Soc. of Am., 94: 3387–3391.

Mei, C.C. 1989. The applied dynamics of ocean surface waves, World Scientific.

Miller, J.G. and D.I. Bolef. 1968. Sensitivity enhancement by the use of acoustic resonators in CW acoustic spectroscopy. J. Appl. Phys., 39: 4589–4593.

Mindlin, R.D. 1951. Influence of Rotary Inertia and Shear on the Flexural Motions of Isotropic Elastic Plates. J. Appl. Mech., 18: 31–38.

Mozhaev V. G. 1989. A new type of acoustic waves in solids due to nonlinearity, Phys. Lett., A139: 333–337.

Murillo, M. and M. Juan. 1995. The 1985 Mexico earthquake, Geofisica Colombiana (Universidad Nacional de Colombia) 3: 5–19.

Naugolnykh, K. and L. Ostrovsky. 1998. Nonlinear wave processes in acoustics, Cambridge University Press, 1998.

Nelson, S.A. 2003. Tulane University, Physical Geology.

Ogilvy, J.A. 1987. Wave scattering from rough surfaces. Rep. Prog. Phys., 50: 1553–1608.

Oldham, R.D. 1900. On the propagation of earthquake motion to great distances. Phil. Trans. R. Soc., A, vol. 194, London.

Pavlenko, O.V. 2007. Self-Modulation of Seismic Waves in the Subsurface Soil (in Russian). Doklady Akademii Nauk, 414: 676–682.

Pao, Y.H., W. Sachse and H. Fukuoka. 1984. Acoustoelasticity and ultrasonic measurements of residual stress. In: Mason (ed.). Physical Acoustics,17: 61–143.

Parker, D.F. 1988. Waveform evolution for nonlinear surface acoustic waves. Int. J. Eng. Sci., 26: 59–75.

Parker, D.F., A.P. Mayer and A.A. Maradudin. 1992. The projection method for nonlinear surface acoustic waves. Wave Motion, 16: 151–162.

Peterson, I. 1986. Mexico City's earthquake: lessons in the ruins, Science News (Society for Science), 129: 36.

Rayleigh, J.W.S. 1885. On waves propagated along the plane surface of an elastic solid. Proc. London Math. Soc., 17: 4–11.

Rischbieter, F. 1965. Messungen an Oberflachenwellen in festen Korpen. Acoustica, 16: 75–83.

Rudenko, O.V. 1983. To the problem of artificial nonlinear media with resonant absorber, Akust. Zh., 20: 398–405.

Reutov, V.P. 1973. Radiophys. Quant. Electron., 16: 1307–1311.

Sabine, P.V.H. 1970. Rayleigh-wave propagation on a periodically roughened surface. Electronics Letters, 6: 149–151.

Sang-Hoon, K. and P.D. Mukunda. 2012. Seismic waveguide of metamaterials. Mod. Phys. Lett. B, 26: 1250101–12501015.

Sato, H. and M.C. Fehler. 1998. Seismic wave propagation and scattering in the heterogeneous earth. Springer-Verlag, New York.

Scherbaum, S., D. Gillard and N. Deichmann. 1991. Slowness power spectrum analysis of the coda composition of two microearthquake clusters in northern Switzerland. Phys. Earth Planet Inter., 67: 137–161.

Sinha, B.K. and K.W. Winkler. 1999. Formation nonlinear constants from sonic measurements at two borehole pressures. Geophysics, 64: 1890–1900.

Slobodnik, A.J. 1978. Materials and their influence on performance, ed. by A.A. Oliner, Springer-Verlag, 225–301.

Snieder, R. 2002. Scattering and inverse scattering. Pike R. and P. Sabatier (eds.). Pure and Applied Science. Academic Press, San Diego, 528–542.

Stoneley, R. 1924a. Elastic waves at the surface of separation between two solids. Proceedings of the Royal Society of London, A106: 416–428.

Stonley, R. 1955b. The propagation of surface elastic waves in a cubic crystal. Proc. Roy. Soc., A232: 447–458.

Stover, C.W. and J.L. Coffman. 1993. Seismicity of the United States, 1568–1989 (Revised), U.S. Geological Survey Professional Paper 1527, United States Government Printing Office, 99: 180–186.

Sumner, D. 2014, Vox, Natural Disasters.

Teasdale, M. 2010. American Geological Union Blogosphere on November 2, 2010 http://blogs.agu.org/landslideblog/2010/11/02/the-canterbury-earthquake-images-of-the-distorted-railway-line/

Tiersten, H.F. 1969. Elastic surface waves guided by thin films. J. Appl. Phys., 40: 770–789.

Tutuncu, A.N., A.L. Podio, A.R. Gregory and M.M. Sharma. 1998a. Nonlinear viscoelastic behavior of sedimentary rocks, part I: Effects of frequency and strain amplitude. Geophysics, 63: 184–194.

Tutuncu, A.N., A.L. Podio, A.R. Gregory and M.M. Sharma. 1998b. Nonlinear viscoelastic behavior of sedimentary rocks, part II: Hysteresis effects and influence of type of fluid on elastic moduli. Geophysics, 63: 195–203.

Torres-Silva, H. and D. Torres Cabezas. 2012. Chiral changes on ratio of the compressional velocity to the shear velocity of earthquakes, International Journal of Research and Reviews in Applied Sciences, 12: 517–522.

Torres-Silva, H. and D. Torres Cabezas. 2013. Chiral Seismic Attenuation with Acoustic Metamaterials. Journal of Electromagnetic Analysis and Applications, 5: 10–15.

Thurston, R.N. and K. Brugger. 1964. Third-order elastic constants and the velocity of small amplitude elastic waves in homogeneously stressed media. Phys. Rev. A, 133: 1604–1610.

Thurston, R. 1966. Wave propagation in fluids and normal solids. *In*: W. Mason (ed.). Physical Acoustics, Vol. 1. Academic Press, New York 1966.

Viktorov, I.A. 1967. Rayleigh and Lamb waves: physical theory and applications. Plenum Press, New York.

V'yun, V.A. and I.B. Yakovkin. 1990. pp. 56–60. *In*: M. Borissov, L. Spassov, Z. Georgiev and I. Avramov (eds.). Surface Waves in Solids and Layered Structures (World Scientific, Singapore).

Whitham, G.B. 1974. Linear and Nonlinear Waves, Wiley, New York.

Zabolotskaya, E.A. 1992. Nonlinear propagation of plane and circular Rayleigh waves in isotropic solids. J. Acoust. Soc. Am., 91: 2569–2575.

Zamora, M. 1990. Experimental study of the effect of geometry of rock porosity on the velocities of elastic waves (in French) Docteur es Sciences Thesis, University of Paris VII.

Zarembo, I.K. and V.A. Krasil'nikov. 1971. Nonlinear phenomena in the propagation of elastic waves in solids. Sov. Phys. Usp., Enlg. Transl., 13: 778–797.

Zhuang, S., G. Ravichandran and D.E. Grady. 2003. An experimental investigation of shock wave propagation in periodically layered composites, Journal of the Mechanics and Physics of Solids, 51: 245–265.

Index